i in the Sky: Visions of the Information

i in the Sky:
Visions of the Information Future

Alison Scammell
Editor

INFORMATION MANAGEMENT

Published by Aslib/IMI

ISBN 0 85142 431 7

© Aslib 1999

Information Management International (IMI) is a trading name of Aslib.

Aslib/IMI provides consultancy and information services, professional development training, conferences, specialist recruitment, Internet products, and publishes primary and secondary journals, conference proceedings, directories and monographs.

Aslib, The Association for Information Management, founded in 1924, is a world class corporate membership organisation with over 2000 members in some 70 countries. Aslib actively promotes best practice in the management of information resources. It lobbies on all aspects of the management of, and legislation concerning, information at local, national and international levels.

Further information is available from:

Aslib/IMI
Staple Hall
Stone House Court
London EC3A 7PB
Tel: +44 (0) 20 7903 0000
Fax: +44 (0) 20 7903 0011
Email: aslib@aslib.co.uk
WWW: http://www.aslib.co.uk/

Contents

Editor's Introduction

i in the Sky: Visions of the Information Future comprises personal predictions of the future by authors representing a wide range of backgrounds and occupations. As well as information professionals and academics there are entrepreneurs, journalists, a 'virtual lawyer' and best-selling management writers. The book even includes a piece of fiction, written by Orange award-nominated novelist, Lise Leroux.

Predicting the future is always a risky occupation and it was not surprising that many potential authors turned down the request to write for the book because they did not wish to associate themselves with the crystal ball gazing fraternity. But people like to read about the future, especially as we enter the new millennium, and in the information business educated guesses on the direction of this most important of commodities is a useful, albeit fraught, activity. The purpose of this book is to provide a space for writers from a diverse range of information backgrounds, to be as uninhibited, visionary and as creative in their forecasts as they like.

In briefing the authors a precise time span for 'the future' was not specified. This ambiguity resulted in a number of different personal interpretations and in many cases a time span was ignored altogether. Others were more precise and at one extreme Richard Wakeford touches on the 'deep future' which he defines as beyond the 100,000 year horizon, while on the other hand, Dave Nicholas and Tom Dobrowolski base their ideas around the concept of 'the future of information is now'. In many ways our definitions of the future are irrelevant as it is probably more important and certainly more interesting to dwell on the significance of information developments, to get at the big picture, rather than focus on the time scales in which they are likely to occur.

The book deals with information in its widest sense. As David Raitt asks, 'does it embrace pretty much everything – remote sensing data, traffic light signals, a sky filled with foreboding cumulonimbus clouds?' Many authors found it a formidable task to discuss a subject which had not been adequately defined. Information, however, like politics, is everywhere, inseparable from other human endeavours. The contributions in this book attest to the penetration of information in all aspects of life. The all-pervasive nature of information is described by Erik Davis who sees it becoming increasingly more invisible and 'more embodied, simply because its flows will literally penetrate the world. Information will no longer lurk in cyberspace, but will be woven throughout our commodities, dwellings, vehicles, even our bodies. Information will be part of the world.'

The ubiquity of the Internet and the increasingly rapid convergence of digital media are discussed in this book along with wearable computers, intelligent household appliances and even computer implants. Professor Warwick's chapter discusses the symbiosis of man and machine, illustrated with a dramatic example from his own experience. In 1998 he had a silicon chip transponder surgically implanted in his left arm, a procedure which was intended to explore the synergy between silicon and carbon, integrating human functions with computer processes. With stories of intelligent pacemakers hitting the headlines and electronic tagging already a well-established fact, this experiment is not as far-fetched as it may at first appear. Indeed, Professor Warwick maintains that before too long thought control of computers will be a realistic development and he includes examples of some recent innovations to back up this vision.

Authors generally agree on the increasing trend of the customisation of information. The ability to personalise data, products and services on the web is re-shaping marketing strategies. These trends are not always seen in a positive light and examples are given throughout the book of information that has been manipulated or customised, at the expense of the truth, for a specific market or audience. One example, regularly cited, is the Encarta encyclopaedia's entry for the inventor of the telephone, given as different people in different language versions.

The shifting demographics and the rise of the e-generation are shaping the future of the web and having a profound influence on how e-commerce evolves. Don Tapscott describes the shift from a broadcast media to a culture characterised by the new one-to-one interactive marketing model. Parallel to the commercial environment, he also refers to the paradigm shift in education where the 'one-size-fits-all, lecture-oriented, teacher-focused approach to instruction is being replaced by customised, interactive, student-focused learning which exploits networks and the new media.'

Other contributors discuss the importance of information in higher education and the implications for the future. Philippa Levy's chapter discusses the networked learning model in the light of a more multidisciplinary and collaborative culture where role convergence (between IT staff and librarians and librarians and teaching staff) is accompanied by a range of technological innovations stimulating and encouraging virtual learning. The importance of embedding information skills and support for the learning process, as a central part of the subject curriculum, is seen as fundamental.

Frank Colson introduces a new concept integral to the future of both academic and commercial enterprises, that of the 'ateliers'. Loosely defined as 'studios housing scholars/artisans, engineers and writers working on cognate subjects whose boundaries are defined by no particular technology…concerned with publishing their work in both digital and print form. Small but ever changing in shape, agile yet robust, the atelier will become a primary agent of change in the information society of the future.' This idea of the atelier harks back to the scholar-printers of the fifteenth and sixteenth centuries. Today's ateliers will herald some far-reaching changes affecting education, artistic and commercial developments along the lines of a new information-driven renaissance. The ateliers will have another fundamental role, concerning authority of documents in the digital era, spurring the development of more powerful multimedia search engines and aiding in the production of high-quality scholarly tools such as thesauri.

This need for authority is highlighted by a number of authors in discussing the issue of information overload. For Gerry McGovern, information overload is 'the single greatest problem not simply facing the Internet but facing all of us in practically every aspect of life in the Digital Age.' Others see information literacy as an essential solution to this problem which needs to be addressed at all levels, starting in the early school years. The emphasis is shifting towards better organisational usability of information systems and individual empowerment. This needs to be accompanied with more sophisticated filters, robots, portals and accreditation and validation mechanisms (whether automated or human). This 'plug and go' environment will gain in importance as more people work for themselves in an enterprise culture which favours information-based occupations and services. To counter the prominence placed on the need for more efficient retrieval tools, Sheila Webber emphasises the need to examine the way in which individuals interact with information rather than the way in which information is used in libraries and organisations.

The key to solving information literacy is seen as cultural as well as technological. We have already seen that the uptake of knowledge management techniques in organisations is only really successful where a major cultural change has taken place. Flatter hierarchies, collaborative and project-based working have been an essential prerequisite to forging a knowledge sharing culture. In discussing the issue of information literacy, Pita Enriquez-Harris also sees the need to ask certain questions about the human brain's capacity to handle information and whether we can ever hope to be better at the job of producing knowledge from information. This may require a certain amount of 'rewiring of our mental processes' to enhance our value judgements and our decision making.

Another solution is provided by Keith Devlin, who, controversially, does not believe there is currently such an occupation as information scientist.

His solution to the growing importance of information is to develop a common sense understanding of it, based on a firm theoretical (and probably mathematical) framework which he has dubbed 'infosense'. He claims that without this we will 'handle information with the same dangerous ignorance … with which Marie Curie handled radioactive substances.'

If information is considered here as a potentially dangerous substance how optimistic are the other authors in their visions of the information future? Kevin McQueen warns about making assumptions that the information future is so bright and urges us to consider the downsides of technology in the light of the political consequences of use of the Internet in China. Lise Leroux's short story paints a menacing vision of human error in a tale of virtual reality. Many authors refrain from commenting on whether or not their predictions indicate a benign or inauspicious world. For others it is simply too early to tell or, as Erik Davis says, 'the jury is out on whether the information age is making us knowledge workers or data slaves.' Perhaps we should bear in mind Charles Handy's comments that although information in the future will be a mixed blessing, we shall adapt and in the end 'life, love and laughter will continue'.

I hope by reading *i in the Sky* you will be better equipped to deal with the information you use in your daily routines, both at work and at home and it will help you make decisions about how you might plan your information use in the future. Above all I hope you will be entertained, enlightened and even surprised by what you read.

Alison Scammell
London, October 1999

1

Darwin Among the Books

Richard Wakeford

1. Visions of the Future

When you are up to your neck in the future – business plans, corporate plans, and next week's rota – it is difficult to find the time to have a Vision of the Future. Some places, however, do have a sense of the Future embedded in the stones, and sitting here at a desk in the new British Library I can, at times, feel borne forward by the flow of literature collected over centuries and by this building, constructed to endure until the middle of the next millennium.

We stumble backwards into the future with few markers and no map to guide us. Companies and large organisations are seldom able to look forward more than a couple of decades and their operational plans rarely extend for more than seven years or so. (Paradoxically, their employees have an eighty year horizon for their own personal finances – but only if they are well advised, and if they are long lived.) Systematic futurological studies have been popular since the 1950s, and although worthwhile as a means of getting people to raise their eyes from the here and now, futurology has had a mixed track record when it comes to the accuracy of its forecasts. Understandably, futurologists are rarely around after twenty years to evaluate their predictions. One such scrutiny, though, was written by Nigel Calder,[1] editor of the *New Scientist*'s 1964 study 'The World in 1984', who revisited its forecasts and concluded that 'the mean rating scarcely reaches "fair"'. The commonest fault was over-optimism about the rate of technological change. Artificial intelligence was flagged by several contributors to become one of the key technologies of the late twentieth century. Now, sixteen years after Calder's scrutiny, we can see that the key technologies that did come to pass (e.g. the personal computer and the Internet) were correctly forseen but their significance and impact were totally missed. Perhaps all those seeking to 'future proof' their company should remember Woody Allen's line, 'How do you make God laugh? Tell him your plans for the future'.

Companies are short-lived phenomena; most companies that were prominent before the Second World War are now no more and few of today's majors will be around to employ our great-great-grandchildren at the turn of the next century. Company names, some with roots in mediaeval times, may persist to give a gloss of permanence but long-standing enterprises such as Beretta, the Italian gun maker, founded 470 years ago and now the world's oldest industrial enterprise, are extremely rare. Beretta is even more exceptional in that after thirteen generations it is still a family enterprise. Contrary to intuition, most family names and enterprises are surprisingly short lived. Studies of groups, such as the English peerage, which have long and comprehensive records, show that the majority of surnames do not survive after two hundred years.[2] With merger, acquisition, or death of the owner, comes loss of institutional memory and vision when the archives follow the corpus of tacit knowledge and tradition out of the door. But old and thriving organisations are not uncommon and it seems that there is something about libraries, universities, churches and their like that confers a long and robust life – only if you are an Oxbridge college or the Church of England do you undertake to release property on a 999 year lease or invest in the cultivation of oak trees. These not-for-profit bodies are immune from takeover or extinction of the family line and serve permanent human needs, such as education and welfare, which do not disappear with the passage of centuries. Only long-lived institutions can undertake long-term projects such as the agricultural field trials that have been underway at Rothamsted Experimental Station (now the Institute for Arable Crops Research) for the last 150 years.[3] These long-established projects produce the data that enable subtle patterns to be detected in noisy environments and increasingly they are being called upon for aid in solving problems in ecology and global management. Although universities may be robust, research programme funding is fragile and many long-term monitoring projects have only produced a few decades worth of data before being terminated.[4] Unfortunately, monitoring is not the sort of activity that is competitive in producing the short-term scientific yield which now must satisfy funding bodies or the career needs of individual scientists. To limit the damage to oceanographic monitoring, the Global Ocean Data Archeology and Rescue project has been scouring archives to secure records, some of which extend back over a century, before they are lost.[5] On a more modest scale of long term scientific enquiry, I am rather taken with the dedication of the University of Queensland to the Pitch Drop Viscosity Experiment which has been running since 1927 – 'the seventh drop fell during World Expo88 and an eighth drop (1995) – is doing well'.[6]

If there is one major challenge posed by the long-term future it must be how we are to manage the disposal of high-level radioactive waste. Plutonium with a half-life of more than 20,000 years must be interred for more than a million years before it can be left in safety. A time span of this magnitude and the unknown risks involved raise questions very different from those with which we are used to dealing. How should engineers cope with the next Ice Age in their design? In this context the next Ice Age is a short-term phenomenon, but it is at least a predictable one as we know from models of the Earth's climate that is highly likely to occur 50-60,000 years from now. How can we communicate with people who will not be able to recognise modern languages and who will have forgotten everything about the purpose of the storage structure? The waste store must endure for many times the life of Stonehenge and that is a monument which after only 4000 years has become an enigma, even though its builders presumably thought its purpose clear enough. A means of communicating across millennia was devised by the US Department of Energy which proposed building a 'Sign of the Ages' – a hazard warning in the form of a menacing earthworks resonant with unconscious messages of fear and dread.[7] A more benign approach to talking to our descendants is the suggestion for a 10,000 year 'Clock of the Long Now' and an associated library.[8] The clock will tick once a year, bong once a century and the cuckoo will come out to mark each millennium. The clock will symbolise the long view of things for contemporary people and the library will archive all important human knowledge to be on hand when the next renaissance is due.

2. Visions of Information

As landmarks for the future are few and unhelpful, is there any framework for understanding how the records of human knowledge will change over time? It seems to me that there is a simple key to understanding and it lies in the fact that information of any sort *replicates* itself. This means that at a fundamental level we can look to see if the logic of Darwinian natural selection governs records in the same way as it does biological systems. The similarity is deepened by the digital nature of the records and the genetic code. The idea that culture and language form a system of extrasomatic inheritance subject to evolution is not a new one and goes back at least to Charles Darwin. More recently, Richard Dawkins introduced the concept of the *meme* which, by analogy with the gene, is a unit of self-replicating cultural evolution, a distinct memory, idea, or behavior, that resides in the mind and spreads between minds.[9] It is when the memes emerge from the mind in physical form that we can look for the Darwinian rules at work. Memes spread either through written records (e.g. chain letters, the propagation of Bibles, or best-selling novels), by speech (e.g. epic poems told by bards, children's playground rhymes or the 'living books' of Ray Bradbury's novel *Fahrenheit 451*), through objects (e.g. the craze for yo-yos, the spread of the wheel in prehistoric times) or by actions (e.g. dance fashions). Memes that employ a digital code, such as alphanumeric text or musical notation, are a special case as they can be copied with very high fidelity in the same way as genes are on replication. Mutations are rare and when they occur they often are selected out of the population – as, for example, the short lived *Wicked Bible* in which the seventh commandment says 'thou shalt commit adultery'[10] and perhaps one way to think about a high fidelity meme is as the basis of a unit of 'copyrightability'. It is easy to take reliable copies for granted but it is a remarkable fact that high fidelity replication creates long lineages in which ancestors and descendants are identical even though separated by many generations. Some genes, exposed to intense selection, have passed down essentially unchanged for millions and even billions of years. In the same way canonical texts such as the Bible or those of classical authors, selected through constant proof reading and criticism, have come down to us after hundreds or thousands of years. Memes that act as sources of textual authority and are supported by high fidelity replication are in a way comparable to the animal germ line, the source of hereditary material that is protected from mutation and kept ready for transmission to the next generation. In contrast memes which are transmitted by low fidelity replication or analogue media such as gossip, visual demonstration, diagrams and sketches or whatever, lack a source of textual authority and are liable to recombine and mutate to create new and stimulating patterns. Memes of all kinds are, however, intimately related and constantly interplay with one another. As the saying goes, 'A scholar is just a library's way of making another library'.

Memes that live in a Darwinian world compete and adapt to their environment. As Daniel Dennett has written, 'The normal state of affairs for any sort of reproducer is one in which more offspring are reproduced in any one generation than will reproduce in the next. In other words it is nearly always crunch time'.[9] The population of texts continually threatens to outstrip the available space on shelves, hard disks or publishers' lists. Selection is intense as, of the 10^5 or so books published in English every year, only a few prove sufficiently adapted to popular interest and fashion to be reprinted or succeed in becoming second editions. Of these only a tiny fraction will eventually become established with a 'classic' status. Immortality for the few is purchased at the price of extinction of the many and, as one biologist put it, 'to a first approximation all species are extinct'. The history of organisms is marked by mass extinctions, periodic storms that leave few survivors, and by background extinction, a slower process of attrition of biodiversity. Both patterns seem to have parallels in the world of high fidelity memes. When the Dark Ages overtook the culture of the classical world, few authors survived and it is a shock to think of the riches that are now lost and were described in the 120 scrolls of the catalogue of the Library of Alexandria.[11] At a day-to-day level the stock of any library is threatened by deletion, physical decay, acts of God, neglect and human destructiveness. Some media, especially magnetic storage, decay after a few years, wood pulp paper fragments after a few centuries, and acid free paper may last for a few millennia.

Libraries are notoriously at risk from fire and unfortunately people are also from time to time prone to consign books to the flames. Texts which are important on first appearance or printed in many copies have no lasting protection against extinction. Already many of NASA's magnetic tape records of the space programme have decayed and are inaccessible. One of the world's all time Top 10 best-sellers, a curious business publication called *Message to Garcia,* with 50 million copies printed since 1899, is today unread and reduced to a few copies held in national libraries. It has, though, been recently revived by publication on the web.[12] Will low cost electronic storage media provide the solution to storing everything in perpetuity? Probably not – the mechanics of archiving are problematic[13] and digital information is being produced at an explosive rate. A recent review has described the pressures facing producers of large scientific databases who 'are already bracing themselves for what they call the exabyte challenge (1,000 petabytes [or 10^{18} bytes]). All the words ever spoken by human beings amount to about 5 exabytes. In case we need them, and we probably will soon, terms have been coined for 1,000 exabytes (zettabytes [or 10^{21} bytes]) and 1,000 zettabytes (or 10^{24} bytes)'.[5] If replication is the criterion of life, then extinction comes soon to most high fidelity memes in the sense that they become unread, uncopied and unloved, but their final physical destruction in the last library of last resort lies in a more distant future.

What is the scale of this future? The past is our only guide. We have inherited our corpus of texts over the last 1500 years or so; urban civilisation has been around for some 10,000 years, culture in the broadest sense has been with us for some 35,000 years and our species is about 100,000 years old. Can we envisage information being transmitted over timescales of the order of 100,000 years? Picture if you will an orderly queue of our descendants: children, grandchildren, great-grandchildren etc. stretching into the distance rather than into the future. Each individual is a child of the one before and a parent to the one after. Allowing 25 years for a generation, then our 4000 lineal descendants would form a line four kilometers long, from, say, the British Library at St. Pancras to (aptly enough) the Natural History Museum at South Kensington. Now the Vision of the Information Future seems to be becoming more tractable. Imagine each parent writing and passing a note to each child with the instruction to copy the note and pass it on down the queue, together with a second instruction to write a new note. The first few of our descendants in the queue cope with the stream of paper but soon bottlenecks develop, pockets and handbags overflow and paper is thrown away. Interesting notes get to be retained and notes of no importance are discarded but this process is haphazard and local decisions are taken regardless of the long range consequences. But the final question of the information future remains – will any high fidelity memes succeed in traveling down to our remotest descendants, or, in time, is all knowledge destined for extinction?

3. Is There an End to Information?

Beyond the 100,000 year horizon the deep future is wrapped in mystery. Maybe the brain and mental functions of mankind will evolve or we may enter into some form of direct symbiosis with machines. If we succeed in achieving sustainability and becoming a very long-lived race we can look forward to an inhabitable Earth for the next few billion years but eventually we will be forced to pack our books and emigrate, for the sun will age and die. Maybe robots will take our place as in John von Neumann's vision of space seeded with self-reproducing automata spreading the message from Earth to the far corners of the galaxy and beyond. Recently our view of the long-term future has become clearer and immortality now no longer seems to be an option. Astronomical observations are coming to support the hypothesis that the universe will expand forever rather than ending in a Big Crunch. The warm phase of the universe, however, will last for only another 10^{12} years. After that the universe becomes increasingly dark and cold as matter is consumed by black holes. Eventually the 'empty era' begins in which black holes evaporate and the last remaining particles decay by around 10^{120} years. Beyond that point physics is silent.[14] If this model is valid then our friendly universe with stars, planets and all the possibilities for life and other complex information rich systems will turn out to have been an interesting but rather temporary affair.

Further reading

1. Calder, N. *1984 and After: Changing Images of the Future*. London: Century, 1983.

2. Raup D.M. and Gould, S.J. *Extinction: Bad Genes or Bad Luck?* Oxford: OUP, 1993.

3. BBSRC Institute for Arable Crops. *The Electronic Rothamsted Archive* http://www.res.bbsrc.ac.uk/era/ (lists papers published on Rothamsted field trails between 1843 to date) (visited 10 July 1999).

4. Duarte, C., Cebrian, J. and Marba, N. 'Uncertainty of detecting sea change.' *Nature* 356 (6366), 1992, p. 190.

5. Reichardt, T. 'It's sink or swim as the tidal wave of data approaches.' *Nature* 399 (6736), 1999, p. 517.

6. Edgeworth, R., Dalton B.J. and Parnell T. *The Pitchdrop Experiment.* 1984
   http://www.physics.uq.edu.au/physics_museum/pitchdrop.shtml (visited 10 July 1999).

7. Benford, G. *Deep Time: How Humanity Communicates Across Millennia.* New York: Avon, 1999.

8. Brand, S. *The Clock of the Long Now*. London: Weidenfeld and Nicholson, 1999. See also: http://www.longnow.com/ (visited 12 July 1999)

9. Dennet, D. *Darwin's Dangerous Idea: Evolution and the Meanings of Life*. London: Alan Lane, 1995.

10. Media Timeline. http://www.mediahistory.com/time/gallery/wickedb.html (visited 10 July 1999).

11. Canfora, L. *The Vanished Library: A Wonder of the Ancient World*. London: Hutchinson, 1989.

12. Hubbard, E. (the Elder). *A Message to Garcia. Being a Preachment*. New York: Roycrofters, 1899. Also available at: http://www.hightechbiz.com/pub/messagetogarcia.htm (visited 10 July 1999).

13. Rothenburg, J. 'Ensuring the longevity of digital documents.' *Scientific American* 272 (1), 1995, pp. 24-29.

14. Adams, F.C. and Laughlin, G. 'The future of the universe.' *Sky and Telescope*, August 1998, pp. 32-39. See also: Davies, P. *The Last Three Minutes*. London: Weidenfeld and Nicholson. 1994.

2

The Man With X-Ray Arms – And Other Skin-Ripping Yarns

K. Warwick

Introduction

I remember, as a child, being fascinated by a film entitled *The Man with X-Ray Eyes*. In it, the main actor, Ray Milland I think, put some drops in his eyes and instantly obtained x-ray vision. After a while the effect of the drops wore off and his sight returned to normal. Later he tried it again and again, until, in true Jekyll and Hyde fashion, the effect no longer went away and he was left with a permanent skeletal view of the world. Whilst such a person would obviously have no problems getting a job in a hospital or airport, their life in general, under such circumstances might be somewhat restricted and rather hazardous.

But might it be possible in the future to feed directly into our human brains such information as x-rays, radar, infrared or ultraviolet? Not as a replacement, as was the case for Ray Milland, for our normal senses, but as an extra, perhaps in an easily switchable form. At present this information is converted into a two-dimensional image and fed onto a screen so that our human brains can understand it. Yet computers, that are supposed to be far less intelligent than ourselves, even now, have no problem directly making sense of the information obtained. Cannot we, as humans, be given a chance of having a go ourselves? Well, maybe we can.

The secret lies on the interface between humans and machine, principally between the human body and computers. Historically the two things have been kept physically remote with communication merely by touch or sound. But things are changing fast.

Wearable Computers

Wearable computing has been seen as a way to make individual people smarter.[1,2] This means to augment everyday wearables, such as clothes, shoes or wristwatches, with some element of computing power. It is claimed[1] that in this way wearables can extend the senses, improve memory and help the individual stay calm.

It appears to be a good idea to make use of everyday items that are worn and to enhance them with some computing power. Why not have medical monitoring information in a watch, glasses with an in built display that only the wearer can see or quite simply a belt which contains a computer?

It is said that we will think differently about ourselves when we have computers in our clothing. Indeed, that may well be true; however, the critical element is that the computer is external to our body, so it doesn't actually change who we are other than merely in terms of our self-perception. In terms of our intelligence or abilities these remain the same as when we make use of a portable computing system, apart perhaps from a small time-saving in use.

It is not known whether or not wearable computers will make an impact on our lives. The commercial possibilities of computing power on board a watch would appear to be reasonably good, although it may all be down to fashion and what is socially acceptable as to how much it all takes off.

Wearables suffer from a number of drawbacks in that they must, almost surely, be very lightweight, easily visible or accessible, fairly low cost and low power. All of these characteristics tend to restrict, at the present time, what is actually achievable. However, where the positives of a wearable device are apparent, such restrictions are lifted, as is the case with a Global Positioning System.

Wearable computers are designed to be with the user at all times and to be part of the everyday life of their wearer. In practice, however, most of the suggested devices so far do not tend to fit that bill. In A.P. Pentland's 'Wearable Intelligence', examples given of wearables include dancing shoes which convert dance steps into music, social wearables which are worn as a necklace and flash to communicate the names of their users, stress monitors, and a video camera fixed to a baseball cap, which, it is claimed, enables the wearer to call themselves a 'cyborg'. It therefore appears to be an interesting area in itself, with commercial potential, but with little to contribute to the interface between silicon and carbon, that is machines and humans.

It is worth stressing that wearable computers, in common with portables and even mainframes are subject to the usual human-machine interface problems. To translate signals from the human brain to machine brain, and vice versa, a laborious process must ensue, involving conversion from electronic signals to mechanical movements and back to electronic signals again. This all takes time and energy, and is prone to error. The true merging of silicon and carbon will need a much closer interface if we are to move forward.

Silicon/Carbon in Humans – Why Not?

Not only have we seen quite a number of science fiction writers pointing to a near future in which humans and technology are inextricably linked[3], in recent years many scientists have also pointed in this direction.[4] Partly this is felt to be due to the unravelling of DNA molecules, possible allowing for computers themselves to be constructed from living DNA, and partly it is felt to be due to the likely acceptance, amongst humans, of the use of implant technology to enhance their capabilities. A key driver amongst this is the potential for linking the human neural system with the artificial version.

But why should we even wish to consider such a move? The answer is simple. Machines are already capable of doing many things that humans themselves cannot do, and in the future their range of abilities will be even greater – in fact making machines more physically capable than ourselves has been something ongoing for several thousand years. In this way we have been able to benefit from the capabilities gained with few associated problems, although in many cases we have had to change considerably the way in which we live in order to incorporate the new-found technological advances, recent examples being the telephone and automobiles.

Occasionally problems occur with technology when we lose control of a situation, e.g. when we drive a car too fast, fly a plane too low or explode a nuclear bomb. However, in each case the consequences are fairly limited and those still alive tend to accept that such problems will happen from time to time.

With machine intelligence the picture is rather more dangerous, something that many people are becoming aware of. In some ways computer-based machines can, even now, easily outperform any human in measures that are regarded as being an indication of intelligence, e.g. mathematical processing, memory or logic. It is also accepted by many as just about inevitable that machines will, one day, exhibit an overall level of intelligence that far exceeds anything humans themselves can offer.[5] One thing that is clear is that when we do switch on a network with an intellect roughly comparable with our own, with sufficient accompanying power, it will most likely not wish to be switched off again.

Interestingly, predictions from the past, both in science fiction[6] and science[7], pointed to the construction of general-purpose robot machines that were humanoid in form and action. What we have witnessed, however, is the appearance of an intelligent, computer-based network, the Internet, linked, to an extent at least, with relatively low intelligence machine/robot entities. Essentially, the robots themselves are not necessarily particularly intelligent, but due to the power of communication, this ability lies in what is controlling them, whether that be human or machine.

The main question about how humans interface with machines is therefore not the one that might have been expected in terms of our interface with individual robots/machines, but rather it is a question of how we interface with networks, and in particular the Internet, a global communications network.

Historically humans and machines have existed as complementary beings, coming together to achieve a particular goal, indeed with wearable computers this is still the case. In relatively few instances only are human and machine inextricably linked, the norm being for physical separation with communication via human external sensing and actuating, such as vision or touch. Contrary to this, in order to help certain humans with particular physical disabilities, the human body has been acceptably breached. Hence we now see replacement hips, cochlea implants and heart pacemakers in a growing number of individuals. For many the alternative is unbearable – quite simply, the implant, with its inherent dangers, must be carried out. Not to do so might mean a life of misery or even no life at all, so the recipient has no choice!

But it is also possible to enhance the capabilities of all humans by the use of implant technology. An implanted individual will be capable of doing things they could not otherwise achieve. However, it may turn out that it is something we will all *have* to do, in order to be part of the system!

Implant Interfaces

In August 1998 I had a silicon chip transponder surgically implanted in my left arm. As I entered the main door to the Cybernetics Department at Reading University, a radio frequency signal across the doorway excited the coil in the transponder, providing current to the silicon chip circuitry, allowing for the repetitive transfer of 64 bits of information. The building's computer was able to recognise me from a unique signal transmitted from my implant. So it welcomed me with 'Hello Professor Warwick', it switched on the foyer light and selected my web page on the video screen. Elsewhere in the building, as I approached my laboratory the door opened automatically, allowing me to pass freely. The computer kept a record of when I entered the building and when I left, which room I was in and when I got there. A location map on the computer gave a real-time picture of my whereabouts at all times. Whilst such an experiment indicates some of the possibilities of intelligent buildings, it also indicates an enhancement in the capabilities of the human body.

Since then a research team at Emory University in Atlanta, Georgia, implanted a similar transmitting device in the brain of a stroke victim, linking up the human neural signals with the silicon present. In this way brain signals were transmitted directly to a computer, and as a result, before long the patient had learnt how to move a cursor around on the computer screen simply by his thoughts, which after all are merely electronic/electrochemical signals. Importantly, it is not necessary for the computer to interpret the signals received but rather for the implanted individual to learn how to send signals to affect the computer in a desired way. The Emory group now intends to use the same device to help paralysed patients turn on light switches and send e-mails. Meanwhile, at the University of Maine, a team led by Ross Davis has transmitted signals from a computer into a patient and, as a result, has been able to control some movements and physical activity.

Technically we are now getting to grips with the fact that signals on the human nervous system can be transmitted to, and received from, a computer, via a relatively straightforward implant. The implant's radio connection with the computer means that a much closer link can exist between human and machine. It should therefore be possible, fairly soon, to interact with and operate computers without recourse to relatively slow computer keyboards or a computer mouse. Indeed, before too long thought control of computers should be a realistic feature.[8]

Linking up the human brain with silicon opens up all sorts of possibilities. Some feel that by enhancing our own network in this way we will be able to effectively cope with information overload.[9] Meanwhile, machines can sense the world in a much richer way than humans, taking in such as radar, infrared and ultraviolet signals, as well as the normal human senses, which in themselves only represent a very small range of the overall spectrum. So it may be that we can harness this for ourselves, directly bringing in such signals onto our nervous system, thereby giving us extra senses. But the whole gambit of machine intelligence also becomes directly accessible, giving us the opportunity of thinking in many dimensions, not just in 3-D or 2-D simplified views, which is what we are used to as humans. Physics books will have to be rewritten when we start to view the world as a multi-dimensional entity.

Linking silicon with the human brain will change the way we think, though, not only in terms of making sense of the new signals arriving but also what we think about ourselves and who we are. In my own implant experiment, after a couple of days I treated the implant as part of me – not an experience, I suspect, shared by wearable computer owners or smart card holders. But due to the things that happened automatically, I quickly developed an affinity to the computer with which 'I' was directly connected. This feeling was a strong one, and not at all what I had expected.

With a close silicon/carbon link, particularly where, via the nervous system, a human brain is linked closely with a machine brain, individual identity will be difficult to define. This is even more pertinent when a human is connected along with many others into an intelligent machine network.

Thought Communication

A link from the human nervous system to computer, via a radio frequency implant, forms a short circuit connection between human neurons and a computer. This opens up possibilities of the computer net becoming an extension of the human brain, with all its memory and mathematical faculties giving the human an extended 'super brain'. Certainly neural movement signals and emotions will be able to be transmitted, and received, and, with a certain amount of learning, various thoughts too.

But once connected in this way to a computer, by means of satellite or the Internet, such signals can be sent around the world. Another implanted person, who can transmit their own signals in return, can receive the signals sent. What this presents therefore is the possibility of global communication by thought processes. Essentially thought to thought communication by humans.

Certainly problems with the technology will have to be overcome. For example, what will one person actually make of a signal transmitted onto their own nervous system from someone else's brain? Will such a signal mean exactly the same thing to different people? The age-old philosophical question, 'When one person thinks of the colour red, is it the same thought in another person?', will perhaps be answered, one way or the other.

But we will need to learn how and when to transmit our thoughts in a way that others will understand. It may in fact be possible for one person to read the mind of another, knowing what that person thinks and what they are about to do, before they do it.

As for the link between human and machine intelligence, it is looking at a future where the two are connected by a rapid radio frequency link. This means that signals from one will be able to bypass the more normal, and relatively slow, sensory systems that exist at present. The operation between the two will be much more in the sense of one overall neural network, part silicon/part carbon.

Conclusions

The whole area of linking human neurons with computers is an extremely exciting one, but as yet is an area in which research is only just opening up. Importantly, it is an area where actual research involving trials and experimentation is required. The impact of the field, in science and philosophy will, however, be immense, as it clearly questions what it means to be an individual human.

Humans still have a lot to learn if they are to fully harness the power of intelligent machines. This is particularly true when those machines have sensors and actuators that outperform anything humans have on offer. Connecting human brains to machines in this way provides an opportunity to enable the two entities to work together, combining the good features of both in a cybernetic whole. Whether it will enable humans to stay in control of the intelligent machines of the future is another question all together! Can't you wait to get yourself an x-ray sense?

References

1. A. P. Pentland. 'Wearable Intelligence.' *Scientific American* 9 (4), 1998, pp. 90-95.

2. S. Mann. 'Wearable Computing: A First Step Toward Personal Imaging.' *Computer* 30 (2), 1997, pp. 25-32.

3. W. Gibson. *Mona Lisa Overdrive*. Voyager, 1995.

4. J. Forsythe. 'Merging Man and Machine.' *Newsweek International*, March 1999, pp. 42-43.

5. K. Warwick. *In the Mind of the Machine*. Arrow, 1998.

6. K. Capek. *Rossum's Universal Robots*. Fr. Borovy, Prague, 1940.

7. J. P. Eckert Jr. 'The Integration of Man and Machine.' *Proceedings of the IRE* 50 (5), 1962, pp. 612-613.

8. K. Warwick. 'Cybernetic Organisms – Our Future.' *Proceedings of the IEEE* 87 (2), 1999, pp. 387-389.

9. P. Cochrane. *Tips for Time Travellers*. McGraw-Hill, 1999.

3

As I See It (The Future of Information)

Charles Handy

The age of information has arrived. It is hard to believe that the web is only five years old, so much is it already part of our language. Who would have thought that grannies would surf the web and would delight in emailing their offspring – no fear now of interrupting their inconvenient moments, and replies are almost guaranteed, even from the most dilatory of grandchildren. Already I get business cards with only an email address and a web site, for some people seem to live in cyberspace, a word itself unknown fifteen years ago. You can make a model of yourself at Gap's web site, dress it in clothes of your choice and turn it around to see yourself retreating, should you want to see such a sight. Truly we now have a digital mail-order catalogue of infinite possibilities.

We also have a worldwide friendship network to join or leave at our choice. Love on the Internet is fancy-free, and risk-free. Adultery without pain or hurt! We can join clubs without the fear of being blackballed, be the sort of person we can only dream of being in the physical world, reinvent our character over and over again, live ten lives in ten days if we so wish. Reincarnation on demand! More than that, we can tell our politicians what we think of them literally as they speak, putting the people truly in power, democracy made real at last. The whole wide world literally at ones fingertips is a wondrous thought, liberating, mind-expanding, exhilarating.

And yet, and yet .... It isn't all good news. For one thing there is too much of it! Already surveys show that the average executive receives nearly 100 emails and over 100 voice-mails every day. 'Our people have stopped thinking,' one top executive complained to me, 'They are too occupied in responding'. Secretaries may be disappearing from the executive suite only to be replaced by a new breed of information gatekeepers. But even they may not be able to keep insidious invaders at bay. A virus wiped out my address file the other day and lost half an essay. And the mail that does get past the gatekeepers, human or electronic, seems to demand an instant response. Friends ring me on the phone to ask if I have received their email of the day before because they had received no reply. Slow down this digital world, I cry, before I jump off in despair.

Then again, neither speed nor quantity is any guarantee of quality, or of truth. The Internet screens out age and gender, which may be politically correct, but if you don't know who is typing or talking, the veracity has to be dubious. My friend wanted a medical definition of death. Instead of looking in the dictionary he posted a note on the web. 'Amazing,' he said, 'within an hour I had ten replies'. 'Were they all the same?' I asked. 'No,' he replied, 'of course not, it's a dicey question.' 'So how do you know which is the best since you know none of the respondents?' Answer came there none! Or take products: you may be able to read and see the items in the catalogue but touch them and feel them, smell them or taste them you cannot. I like to press my avocados before I buy them to make sure that they are ripe, now I have to trust the store's assertion.

Trust! Ah yes, this new world depends on trust for how else can you rely on people when they are usually out of your sight with an address that locates them only in cyberspace and not in any geographical area? Virtual organisations are wonderfully cheap – they need no offices, heating or cleaners. Strangely, however, their costs of travel seem to escalate, because in the end you cannot rely on people whom you have never met if the problem is outside the normal routines. Already we seem to trust brands more than individuals, because we don't get to know individuals well enough. Europeans love their mobile phones because at least there seems to be a human at the other end. But even mobile phones herald a change in the way we organise ourselves because now a phone belongs to a person not a place. Motorola's alleged vision of a world where every child gets a name and a phone number at birth is not that far off. My niece's four week old daughter already has an email address and will have her own phone number too as soon as she can talk.

Now that those phones can get and receive emails while we walk down the street, who knows where anyone is anymore? But how can you control people if you can't ever know where they are or what they are doing? Offices used to be some sort of corrals for people who had sold their time to the organisation. Now the horses are all loose and there may not be enough cowboys to herd them up when needed. Schools won't, in theory, need to corral their pupils every day, teaching them over the Internet instead, and one can imagine governments attracted by the sort of savings possible with virtual schools. But teenagers aren't all eager students of the Open University, self-disciplined and self-organising. No longer will weary parents heave a sigh of relief when their kids are safely parked. Will we therefore see electronic tagging replacing the roll-call in the schoolroom? And will that be a good thing?

Property, too, becomes baffling. In this new world, ideas, information and intelligence are the new sources of wealth. But this wealth is different. I can give you all that I know, but I still know it and have it, unlike land or cash. Intelligence, likewise, is hard to pin down or stake, unlike land. We can't dish it out or redistribute it, nor can we tax it because what can't be measured can't be taxed. Sometimes we want everyone to know our ideas, but sometimes we want to keep them to ourselves, but how do you patent ideas unless they have a visible shape and form? Increasingly it is going to be harder to own what we produce and that will give lawyers a lot of fun, and much profit. Yet, in some ways, a world of unowned property might boost economies.

Some say, in fact, that an almost-free world of information will bring equality of opportunity to all. The politicians' dream come true by accident of technology. Villagers in India will have access to the world outside no different from the rich man in his Californian hideaway. Knowledge can rain down upon the poor and the wealthy, the near and the far, alike. Education for all becomes a real possibility. Others, however, fear that this new resource of information will, like all the sources of wealth that have gone before, sort out the rich from the poor. Only the rich can afford to buy those portals which are the entry to the web, so that when you seek financial information on Netscape you are confronted with the array of Citibank's products and the lazy will go no farther. And they say that Citibank paid $40 million to get the jump start on their competitors.

Experts think that, before long, 80% of online commerce will be done by just 30 companies. The rich will have hogged it all. It is just that the new rich may be different from the old rich, as has always happened in revolutions, be they of arms or of technology. In which case we shall have to wait a generation or two, no doubt, before the new rich begin to take on the lessons of noblesse oblige and start to do what they can to help the new poor.

Yes, it will all be a mixed blessing, that seems sure. But manna from heaven can't be returned just because it doesn't fall equally or because you don't like the taste. We shall have to learn to embrace the inevitable, not ignore it. In the end we shall adapt, as humans always have, and in the end life, love and laughter will continue, even if the paraphernalia is more exotic and more digital than we have been used to. Spring will still smell as nice, perhaps nicer, since information does less harm to our environment than messy steel or cars, and Shakespeare's plays will still have meaning because they deal with love and jealousy, ambition and avarice, pride and compassion, death and the meaning of life, and such things do not go away. Perhaps, however, more people will see them, in more languages and in more settings and if that spreads a better understanding of life's mystery, that cannot be bad.

4

Maintaining a Balance?

Christopher Davis

The fifty years since World War II have witnessed a third industrial revolution[1] in the economically-advanced countries and a gradual transition to an Information Age or Post-Industrial Era which holds out the promise of an exciting new world of unrestricted freedom of choice, high-tech automated production and material abundance. Alvin Toffler went so far as to label this transition 'the Third Wave', suggesting that it will ultimately be as significant as the two previous waves in human history: from hunter-gatherer to agricultural societies and then from agricultural to industrial.[2]

In this information society, instead of working in a shipyard or a coal mine, the typical worker has a job in a software company, an insurance company or restaurant. The key factors of production shift from materials, energy and labour to information and communication. Production becomes globalised as inexpensive information technology increasingly eases the movement of information across national boundaries and rapid communication via the Internet, television, fax and radio erodes the barriers of long-established cultural communities.

The future of information is not therefore just a technological issue; it also affects the political and socio-economic life of countries. Openness of information has achieved exactly what the communist regimes feared most, the demand by citizens for freedom, equality and participation in the decision making process. Hierarchies of all sorts, whether political, social or corporate, are coming under pressure and are crumbling. Freedom of choice has exploded, leading to a culture of intensive individualism that has spilled over into the realm of social norms, where it has eroded trust and confidence in virtually all forms of authority and has weakened the bonds holding families, corporations, neighbourhoods and nations together.

The cultural theme of liberating the individual from unnecessary and stifling social norms and moral rules that has swept many societies since the 1950s has lead to increasing levels of social dysfunction including the growth of crime, broken families, drug use, suicide, litigation, tax evasion and the alienation of neighbours from one another. Decentralisation and deconstruction is rampant, not just in government, but also in the commercial sector. The recession that affected many countries during the 1980s and 1990s led to a shift in the relationship paradigm between large-scale employers and their workers so that workers no longer believe they can look to employers for secure lifelong employment. Even Japanese companies, perhaps the most paternalistic, are now succumbing to this new reality. This drive to reduce fixed overheads has led to new ways of working, including teleworking and contingent working, where the worker is brought in only as required. In this new hard-nosed environment, workers no longer trust employers to look after them, but instead are looking out for number one even at senior levels, protecting themselves by taking on portfolios of interests.

As the State and employers shed their paternalistic role and forced self-sufficiency on to individuals, fundamental shifts are occurring in the social contract and employment relationships. Citizens and workers no longer pledge such strong allegiance to their country or employer, but engage in mutual benefit in selected areas – closely costed to reflect each party's deposits and withdrawals. The large-scale pooling of resources for the common good, where individuals may or may not benefit from their input, is no longer attractive. However, no man is an island and people will be forced to organise into communities of self-interest, as they did before the emergence of nations and large-scale employers.

Our fundamental need to belong to a community has left many people feeling lonely and disorientated. A new web of relationship formats is already arising to fill the void created by the dislocation of the individual from the State, employer, church, neighbours and family. Many of us have experienced the revival of remote familial and other relationships through email, reflecting how communities are becoming increasingly temporal rather than spatial based on the use of telecommunications and information technology. Perhaps even more significantly, it would appear that a veritable 'association revolution' is underway at a global level as significant a social and political development as the rise of the nation-state in the 19th century.[3] There has been a massive jump in non-governmental organisations such as charities and associations in many countries including the United States, Europe and even the Third World. In summary, it would appear that information will increasingly flow via different human channels in the world of the 21st century, with new types of community, informal norms and greater self-governance.

The increased exchange of information is shrinking the global village to the point that national cultural boundaries are being eroded. The world community is better informed about, and less tolerant of, bad behaviour by its neighbours (witness NATO's recent challenge to Yugoslavia's internal activities), and this is likely to lead eventually to a world made up of many localised geographic and temporal communities governed by one global government and one police force.

These are some random thoughts on the way in which the Information Age is likely to affect society in the future, but what are some of the indicators of future trends that are beginning to emerge with regard to the way humans will use information in the future? My personal list is:

- Wider dissemination of information
- The rise of artificial intelligence
- Improved knowledge management
- Increased remote communication
- Deconstruction of the corporation and rise of flexible working

Wider Dissemination of Information

We are witnessing a second quantum leap in the dissemination of data, information and knowledge[4] to a wide range of the world's population, following on from the invention of the printing press in the 15th century. Not only has there been an explosion in the amount of data and information made available in recent years on a massive variety of subjects, but it is now easier and inexpensive to access, removing the barriers to people doing so.

This information explosion will escalate as our connectivity to the Internet becomes closer; many of us have access to it on our PCs at home and at the office and it is now also available on our televisions. Once this becomes commonplace, we are likely to use it more and more as we relax in our living rooms. The appeal of online access will escalate as the organisation and content of online information improves; for instance, with the recent onset of digital television we are increasingly able to watch and interact with an almost infinite range of programmes. Furthermore, as data is standardised into digital format, every form of data will be delivered to us in compatible format. With the advent of digital television, DVD for video, digital music format on the Internet, FM radio, and digital text, we will be able to receive all data, manipulate it, record it and transmit it from one piece of equipment. The entire world will be delivered to us wherever we are, as long as we have access to a screen.

However, biologically, we humans can only cope with so much of this glut of information, so information portals are having to be developed to filter the information to our personal and corporate requirements. Examples are the news sites on the Internet that allow us to tailor our daily news requirements to suit our personal tastes. Eventually, each of us is likely to have our own portals for all of our information needs, as we filter the vast array of television and film channels, radio channels and Internet sites. Intelligent agents will search and sift the information providers' offerings for us, giving us a taste of what is on offer so we can chose whether we wish to hear more of that musical group, see more of that new television show or read more of that news article.

The Rise of Artificial Intelligence

As Jeremy Rifkin observed in his book *The End of Work*[5], 'Numerically controlled robots and advanced computers and software are invading the last remaining human sphere – the realm of the mind. Properly programmed, these new "thinking machines" are increasingly capable of performing conceptual, managerial, and administrative functions and of coordinating the flow of production, from extraction of raw materials to the marketing and distribution of final goods and services'.

Computer scientists believe that, by sometime during the 21st century, artificial intelligence will be able to out-think the average human mind. They also look to the day when intelligent machines will be sophisticated enough to evolve on their own, creating their own consciousness. The Japanese government has launched a 10 year research project to develop computers that can mimic the most subtle functions of the human brain, including the kind of intuitive thinking that people use when making decisions. They hope to create a new generation of intelligent machines that can read text, understand complex speech, interpret facial gestures and expressions and even anticipate behaviour. Companies such as BBN Systems and Dragon Systems have developed intelligent machines equipped with rudimentary speech recognition which can carry on meaningful conversations and even solicit additional information upon which to make decisions, provide recommendations and answer questions.

All industrial sectors are being transformed by the information revolution. Farm mechanisation, which began with the horse-drawn steel plow in the 1850s, is nearing completion today with the introduction of sophisticated computerised robots in the fields. Man's control of information now even extends to life itself, with the controversial subject of genetically modified crops.

Manufacturing has also undergone dramatic breakthroughs in re-engineering through the continuous process technologies. The idea of automatic machinery producing goods with little or no human input is no longer a utopian dream. For instance, at the Victor Company in Japan, automated vehicles deliver camcorder components and materials to 64 robots which perform 150 different assembly and inspection tasks. Only two humans are present on the factory floor. In 1991, the International Federation of Robotics reported that the world's robot population stood at 630,000. That number is expected to rise dramatically in the coming decades as thinking machines become far more intelligent, versatile and flexible.

Sweeping reforms will also continue to redefine every aspect of white collar and service work as computers increasingly understand speech, read script and perform tasks previously carried out by human beings. The computerisation and automation of the service sector has barely begun but already routine personal services and increasing numbers of more complex service functions are being taken over by intelligent machines, particularly in the banking and insurance sectors. At Cleveland's Society National Bank in the United States, more than 70 per cent of customer's service calls are handled by a voice-mail system. Retailers are also re-engineering their businesses using information technology with automated warehousing and electronic scanning at checkout. A large European super-discounter is experimenting with a new electronic technology that allows the customer to insert his credit card into a slot on the shelf holding the desired product. There is no shopping trolley but, instead, the customer finds the purchased items already packaged and waiting for him when he leaves the store. He simply signs a prepared credit slip and leaves without ever having his items rung up on a cash register. Further, remote shopping will replace personal shopping increasingly, growing at over 20 per cent per year, reflecting peoples' reduced leisure time to travel to shops. Technology already exists for clothing to be displayed on the body electronically, avoiding the need to try-on clothing. Although this service currently requires attendance at the shop, this is likely to be available soon on the Internet.

Improved Knowledge Management

In the Information Age, knowledge rather than physical assets or resources is the key to competitiveness in all sectors. A recent survey by PricewaterhouseCoopers in conjunction with the World Economic Forum found that 95 per cent of chief executives saw knowledge management as an essential ingredient for the success of their companies. Organisations cannot afford to ignore the value they have invested in their knowledge assets.

What is new about attitudes to knowledge is the recognition of the need to harness, manage and use it like any other asset. Significantly, such intellectual capital is increasingly being valued and reported upon by companies.[6]

At its most primitive, knowledge management can be simple as noting contact telephone numbers, photocopying the list and sending it to everyone who might need it. At its most advanced, knowledge management encodes the unencodable; it tries to capture the unwritten tricks of the trade which make an organisation function, store them formally in a database and make them available to employees to improve their performance. This can be taken even further by building intelligent agents into the system to steer a worker in their decisions and workflow. These agents monitor each worker's activities, analyse and compare them with the database, and then make suggestions as to how to improve effectiveness. Our initiative as individuals will be enhanced by the organisation's knowledge pool.

Intranets

The sort of knowledge management systems discussed above are at an early stage of development and many readers will have experienced the first generation of systems through their employer's in-house intranet. For instance, larger organisations are now commissioning from law firms such as ours a legal site on their intranet, where managers can access guidance and precedents as well as liaise on legal issues across the organisation. Thus, a manager in Chile can be applying the same environmental management standards as his colleagues in Poland. They can be using the same standard form of supply agreement, conformed to local conditions. In this way, an organisation can deliver more legal information to its personnel at a lower cost, as managers first check the site for information before contacting the internal or external lawyer. By building intelligent agents into the system, an organisation can monitor the activities of its personnel to optimise its legal affairs. Similar to Microsoft's Office Assistant, a 'desktop lawyer' might spring into life when it senses the creation of legal relations in written correspondence, suggesting issues that might be considered. Compliance with regulatory requirements might be built into systems to reduce the effects of the sorts of disasters that befell the UK investment bank Barings as a result of its trader Nick Leeson.

Extranets

Such knowledge management systems can be expanded to include external parties, thus becoming known as extranets. For instance, my law firm, Davis & Co, has developed systems for mergers and acquisitions to manage the large quantities of data being extracted from the acquisition target by the purchaser's various internal and external due diligence teams. Traditionally, all due diligence data has been provided in paper form, which is difficult to retrieve and slow and expensive to distribute. Davis & Co's extranet stores all the data electronically, and distributes it on CD-ROM and online. All members of the purchaser's acquisition team, both internal and external, therefore share the same data, rather than working in isolation on incomplete material. The data is easily retrievable from an individual's computer, instead of sitting in bundles on the floor. It is also portable – imagine the benefits of having access to all the disclosed documents at a negotiation meeting in Paris while being provided with the latest documentation being scanned in different locations such as Sydney and Los Angeles.

How Far?

How far can this be taken? In theory all of the written, spoken (both online and offline) and physical communication in the office could be monitored, analysed and fed back into the system. This process is already underway as the all seeing eye of closed circuit television peeps into more corners of our lives and State and private organisations collect more and more knowledge about us from the information we inadvertently reveal about ourselves each time we use our credit cards, apply for credit, insurance or a job, respond to a survey or register a newly purchased product. In the 1998 movie, *Enemy of the State*, it was even suggested that the United States security forces monitor all traffic over the Internet for such keywords as 'bomb'. George Orwell's 'Big Brother' is watching and can be expected to pervade our lives increasingly as we become more connected.

More Remote Communication

Greater Mobility

Alexander Graham Bell's invention is only now, after some 120 years, being fully realised. In recent years, the giant leaps forward in telecommunications and the related lowering of costs are giving us the ability to readily communicate with one another wherever we are in the world whether by voice, on video, or in writing transmitted electronically. Without realising it, we are becoming cyborgs like the Six Million Dollar Man of the 1970's and Robocop of the 1980's. We have the technology, but how far will we go to use it? The military are already well developed in supplementing the information flow to soldiers and fighter pilots with sophisticated helmet and other devices. The civilian population have not however been forgotten as we clip onto our belt or wrist a handheld communication device, which in some cases has a startling resemblance to the flip-open Star Trek communicator of the 1960's. We extend our memory banks by carrying electronic organisers, which can automatically arrange meetings for us. Some of us are connected to a central knowledge management or other system that monitors our workflow and analyses our inputs and outputs to improve our efficiency and effectiveness. As a lawyer working on international mergers and acquisitions, around a third of my luggage on trips now comprises the equipment necessary to sustain me out of the office – laptop, printer, mobile telephone and a host of cables and conversion plugs. I carry all of my past and present files with me on laptop so that I can discuss any matter with a client at any time wherever

I am. By connecting the laptop to a mobile phone or landline, I can then transmit a document to anyone anywhere in the world by email.

This sort of mobility means that people can operate from anywhere at any time. They become mobile workers who can operate efficiently and effectively on a global basis without the need of an office. As business becomes more globally mobile, so will its workers. By compressing time and collapsing space, the new electronic wizardry has transformed the very idea of an office from a spatial to a temporal concept.

However, this mobility comes at the price of having to carry all this equipment and being able to connect to the local telephone and electrical sockets. How long will it be before we tire of carrying all this kit and elect for surgical implants?

Sharing Information Electronically

Interacting remotely can be enhanced by the creation of a private online site (such as an intranet or extranet) where a group of people can communicate with one another electronically in writing. For instance, Davis & Co's extranet for mergers and acquisitions is used where a range of in-house people, such as the client and its advisers, need to be coordinated and be able to reach one another wherever they are in the world at any time. Project management tasks can be posted daily for everyone on the team to see, even though they are en route from a meeting in Beijing to Jakarta. Furthermore, these systems do not simply offer one-way traffic. Online discussion forums allow groups of people to interact electronically on specific topics, using group messages. This offers the same sort of spontaneous communication as a round-table meeting, even though people are geographically unable to get together, and, encourages genuine multi-disciplinary co-operation. Clients no longer have to wonder whether all the issues have been thoroughly canvassed collectively by their advisers – they can monitor activity by the hour. Whiteboarding allows lawyers such as myself to discuss and amend a document online with one or more people also able to see it on their screen.

The Future of Remote Communication

The convenience and cost savings of remote communication make this an inevitable trend, with huge potential for improvement in the experience. Videoconferencing is now established technology but its use is limited largely to small sections of the business community and is still fettered by the limited bandwidths of so many telephone networks. Eventually, this is likely to be enhanced or replaced by three dimensional images where one can don a small headset and be present virtually in the remote location, able to see around the room (via cameras) and be seen and heard through a holographic image and speakers. By the end of the first half of the next century, scientists believe it will be possible to create life-size holographic images of computer-generated human beings capable of interacting with real human beings in real time and space. Kurzweil Applied Intelligence believe these three-dimensional images will be so lifelike that they will be indistinguishable from real people.

Deconstruction of the Corporation and the Rise of Flexible Working

Drivers Towards Flexible Working

The technology that enables remote and online communication is being applied by many larger organisations to change the way they work and for new organisations to grow by applying virtual office and contingent workforce techniques. Cost cutting in many companies has led them to the savings available from encouraging many of their workers to work remotely, with such major organisations as the UK communications company BT leading the way.

The labour market is also driving this movement toward more flexible working, in order to avoid the need for travel and to be close to their families. For instance, this demand is forcing the major law firms in the UK increasingly to offer their senior lawyers the ability to operate from a private office at their home. High work pressure in the legal profession means long hours, and this creates tension between business obligations and private life. Through the close proximity of one's family and work, and greater control over working time, a flexible worker increases his flexibility to meet the demands of both and enhances his standard of living in the process.

These changes, combined with the desire of companies to maintain a contingent workforce that can be increased and reduced according to demand, is deconstructing many corporations and leading to new business models.

Future Business Models – A Case Study

An example of one of these new models is my own firm, Davis & Co, which was formed in 1993 around the concept of the use of telecommunications, flexible working, lean management and contingent working.

With the increasing concern of clients at the high cost of legal services, the firm looked at ways to streamline its operation. Expensive City premises to house its personnel were seen as unnecessary in an environment where technology liberated its people from traditional work practices. From the beginning, the founders of Davis & Co introduced flexible working for *all* its 40 personnel. A City office is only maintained for meeting clients, while personnel operate from private offices at their homes or elsewhere.

The comprehensive application of communications technology allows Davis & Co to minimise its need for expensive City office accommodation and other high overheads. The resulting savings exceed 17 per cent of turnover per annum. These savings, in turn, are shared with clients as lower fees.

Clients and solicitors are linked through a secure email system, BT's audio conferencing facilities, facsimile and voice-mail, enabling messages and documents to be exchanged back and forth. The geographical spread of the firm's solicitors is not an issue for clients because incoming calls are patched to them wherever they are in the world.

Davis & Co has also adopted the contingent workforce model, with many of its fee earners and support staff being brought in as required. In this way, seasoned virtual teams of leading lawyers from a wide range of disciplines can be brought into a project at short notice. A lean management style has also been adopted, whereby fee earners manage their own portfolio of clients and support staff are measured by objectives rather than being managed 'by walking around'. Recognising that departments cause divisions and borders that inevitably slow down the decision-making process, the firm operates a flat organisational structure, transferring decision-making closer to the client at fee-earner level.

This lean style of operation, with a small hub and a network of fairly autonomous fee earners and support staff around it is seen by many as a model that is likely to become increasingly adopted by other organisations in the future, particularly in the knowledge intensive sectors where professional training ensures high quality standards at an individual level.

Conclusion

The US Institute of the Future has foreseen that we are going to become paperless in the same way we became horseless; horses are still around, but in industrialised countries they are just ridden by hobbyists. Will digitalisation become so pervasive that it will eradicate other mediums of transmitting information such as the painted picture, clay sculpture and the paper book? A brave new information world awaits us, but we humans are unlikely to stray too far from our need for a balanced lifestyle combining high-tech and low tech, technology and nature, sophistication and simplicity.

References

1 See Jeremy Rifkin, *The End of Work*. GP Putnam's Sons, New York, 1995. Rifkin postulates that the Second Industrial Revolution occurred between 1860 and World War I, when oil began to compete with coal and electricity and was effectively harnessed for the first time, following on from the First Industrial Revolution in the 18th century which harnessed steam and the large scale manufacture of iron.

2 Alvin Toffler, *The Third Wave*. William Morrow, New York, 1980

3 Lester M Salaman and Helmut K Anheier, *The Emerging Sector, An Overview*. Johns Hopkins Institute for Policy Studies, Baltimore, 1994

4 I have differentiated between each of these three terms, treating data as the raw unit of information that requires some form of processing (such as organising) to make it useful, information as data that has been processed so that it is intelligible but not yet productive, knowledge as information put to productive use.

5 Rifkin op. cit. p.60

6 The valuation of intellectual property is discussed in *Due Diligence Law and Practice* by Christopher Davis, Sweet & Maxwell, London, chapter 18 contributed by Kelvin King of Corporate Valuations, London.

5

No Limits

David Raitt

When asked to write about the future of information, with no definition of the term information provided and no indication about how far into the future one should project, then the sky is really the limit. There is so much one could say, so many possibilities to consider, so many could be almost fantasies to imagine. Is the future five years or ten years or twenty or fifty or more? Does information mean the more tangible facts and figures and data stored in libraries, archives and on the web, or does it embrace pretty much anything – remote sensing data, traffic light signals, a sky filled with dark foreboding cumulonimbus clouds for instance, or some wildly gesticulating man at the race track or stock exchange? With half the world (OK, maybe a slight exaggeration, but you know what I mean) now being classed as information or knowledge workers – then is virtually everything we see or hear or write or need or use considered information?

In some respects, these are rhetorical questions and, in fact, what I want to do first is to go back to some earlier works where similar questions regarding the future of information were posed and answered, albeit implicitly. In many of my papers and articles over the years, I have described various new hot technologies and tried to show how these could possibly be implemented in libraries and information services even though this market was not necessarily foreseen for the technologies in question. It just required a leap of imagination and a certain awareness.

In a paper, given at a conference in South Africa fifteen years ago, entitled 'Look – no paper! The library of tomorrow',[1] I envisaged how the library of the future might look – crammed full of the latest voice-recognition, holographic, robotic, flat-screened, intelligent devices and services. And one only has to look around to see if all these things are already operational in libraries and if not, why not? What happened along the way? Why aren't many of these technologies prevalent yet even outside the library world? I know it does require a long time for technologies to be taken up – look at television, even look at the personal computer (outside the USA at least), but even so.

Ten years or so ago I was writing all about electronic books on smart cards – the future of information was surely here and I wanted people (the library and information community) to be aware of what was available. Big companies were involved, kiosks could be set up, books could be printed on-demand. What happened? Why was there a hiatus? (While we're at it – why did CD-I/DVI become obsolete so quickly?) Why is it only today we are seeing headlines in the computer press like 'Electronic books are poised to become a key medium' or 'E-books open a new chapter'? Will they really? Or will we ask in another ten years time – whatever happened to electronic books again?

There were other visionary developments. Over twenty years ago, the Architecture Machine Group at MIT was looking at a new concept of information retrieval where one accesses a data item by going to where it was rather than by referencing it by name. This Spatial Data Management System (SDMS) permitted one to find items on the basis of a more or less definite sense of their location in space, which may be actually present or remembered. The Group had an experimental version of SDMS – a multimedia information room with touch sensitive TV monitors, microphones, pressure sensitive joysticks, data entry tablets and more. The spatial world of SDMS consisted of a single plane or surface called Dataland which was continually visible to the user in an aerial, top-view display. The user could – and I always remember this phrase – helicopter over the surface or landscape and zoom in on any aspect – a map of the world, then a country, then a town, a street, a house, a room, the desk in the room, and anything on it. Navigation by speech was possible and travel by eye-tracking was under development. Promising, you might think. This is the future of information, you might think. But where is it now? What happened to Dataland and spatial data management – do a search on the web and you won't find much at all. Did it just prove a point – that such systems could be created? And once proved, then it was on to the next project or challenge? Didn't anyone want to try and put some such system in operation? Was it just too costly or time-consuming or impractical or slow or was it just overtaken by other technologies or events? We do have active desktops on our PCs nowadays, also voice input and output and handwriting pads, and we do have zoom in maps on the web and in our cars which can show you streets etc. – but nothing quite like Dataland (except maybe in military and covert intelligence contexts).

In 1985, a book by Stan Lee was published called *Dunn's Conundrum*. It was a techno-thriller and I bought it at the time because I had read about the MIT Dataland and I was interested in new technologies and in seeing whether and how they could be applied in library environments. When asked to write this chapter, I did check to see whether I still had it, but sadly no and it is now out of print– though I have just come across a web site offering second-hand copies of it. I can't really recall the actual story, but that's not the point – the hero (Dunn, I suppose) had access to some fantastic information system similar to Dataland. Ask whatever you want and the answer (analysis, correlation, overview, whatever) comes right out – visually, textually, orally, graphically. And there have been countless (OK, at least a few) other books and films which have similar advanced information/knowledge systems. Take *2001: A Space Odyssey* (who could forget HAL?) or *Demon Seed*. Where are they now? Why didn't such ideas and things take off? Wasn't this the future of information? Some technologies did eventually take off, of course, when pulled by the right markets at the right times – we now have data mining for example (on reflection this is what Dunn's information system was clearly using) – but not many of the technology and software are that ubiquitous and in everyday common or garden use.

It was in 1972 that Georges Anderla, a visionary if ever there was, completed a report for OECD assessing information trends over the next fifteen years (from 1970-1985) – in other words, looking at the future of information. At the International Online Information Meeting in 1985, as Conference Chairman I invited him back to review which of his predictions had actually come to pass. His paper gave a critical, retrospective overview of the period 1970-1985 as well as some key guidelines for strategic planning for the next fifteen years (1985-2000).[2] In his retrospective overview he noted that not all of his early predictions proved correct, in fact far from it! He overemphasised mainframe computers, for instance, to the detriment of minicomputers (and where are they now that we all have personal computers?).

And as for his second attempt to envision the future, he felt there was a tendency towards and a growing demand for supercomputers (these are still rather uncommon in most circles I think). He thought there was a considerable pent-up demand for multi-purpose, especially multimedia processing (yes!) and he believed that all-embracing, full-time integration (of networks and office functions in workstations) was a losing proposition – modular systems, connectable at will at a moderate extra charge, were bound to win on cost/efficiency grounds. He believed that videotex (the exciting new bells and whistles system at the time) could, contrary to expectations, never conquer the market in the way TV did since the latter was an entertainment system and its information function was incidental. On the other hand, videotex was primarily seen as an information medium and all other facilities (such as remote shopping and banking) were secondary.

He recognised that the business market was important and thought the best prospects lay in sectors with a low penetration of computer processing (e.g. travel, the arts, museums, leisure activities etc) whose hunger for images, sound, animation, live scenery, could not formerly be satisfied. He also wrote that the clamour for user-friendly interfaces should not be accepted at face value: why waste time and money on such niceties? Of course, the infrastructure for the Internet and World Wide Web was not in place fifteen years ago – although he was correct in at least some of his assumptions (the business market, the way travel, leisure and arts sites have made a strong presence – albeit on the web), though not in others (home shopping and e-commerce are taking off, and the desirability of simple, common, user-friendly interfaces, if not paramount, is at least important).

My purpose is not to criticize or point out Anderla's (or anyone else's) mistaken assumptions or predictions or descriptions of information and knowledge systems made so long ago which didn't come to pass, but rather to note that, with the benefit of hindsight, by looking at some of these older developments, imaginative scenarios, predictions, you can say what you want about how the world might look in the years to come and what information and knowledge-based systems or new information technologies might be around, but it is very hard to be accurate and speculation won't necessarily get you anywhere, and in any event it will all take a very long time for any of the predictions to actually happen anyway.

When looking for the proceedings containing the Anderla reference, I spotted a book in my cupboard – *Libraries and the Future: Essays on the Library in the Twenty-First Century*.[3] I'd forgotten all about it – Wilf Lancaster had asked several authors six or seven years ago to present personal visions on what the library might look like in 25 or 30 years (i.e. around 2020) – how library and information services might differ from those of the day and how we might get from where we were then to that place in the future. The chapters, including my own,[4] make interesting reading in retrospect. How many of us will be right in our scenarios? How many of us are even on the right path just a quarter of the way through the period? The Internet is not mentioned, the World Wide Web, only a year away, is not mentioned (the nearest one gets to it is the Virtual Information Centre), and has anyone heard anything further on teleports these days?

Also quite a few years ago, I was interested in smart houses – the technologies that new homes would have: solar panels, voice-activated appliances, computer-controlled services, infrared connections and so on – and in seeing how such technologies could possibly be applied to libraries and information centres. But at the time, apart from the odd display house, the technologies didn't catch on in any big way. I'd forgotten it then, but I'm remembering it now – the book *Demon Seed* by Dean Koontz first came out around 1973 – and here, again if I recall correctly, quite a lot of the systems in the house were computer controlled – the doors, the blinds, the lights and utilities etc. And this was 25 years ago. And yet on the CNN web page the other day I came across a news story entitled 'Japanese homes for the millennium: little space, but lots of brains'. There is a already a model house it seems – a Home Information Infrastructure called Warp House HII. The system comprises a kind of central nervous system, a depository for reams of information, and the main relay station for data flowing between the house and the outside. Cameras and monitors keep track of who is doing what where and first aid kits link to doctors who can diagnose your illnesses. The Interactive Communication Refrigerator keeps a track of its contents and if you are in the supermarket you can check with your fridge to see whether you need to buy milk, for example.

In the UK, ICL and Electrolux are marketing a fridge incorporating a touch-screen PC, TV, modem and bar-code scanner. Other developmental systems can take this kind of thing a stage further – supermarkets like Tesco and Safeway. Using smart customer loyalty cards, supermarkets can analyse up to four months of purchases for a given shopper and come up with a definitive list from which the shopper can select items using a PalmPilot and then couple it to a mobile phone and send to the shop. There have been vending machines available for a year or so, for instance, which sense when they are out of one particular item and then automatically dial the supplier to place an order for more. Scary? No – the future of information. Information not necessarily as we librarian and information specialists know it, but information nevertheless. But look at the time it has taken from Koontz's book (and probably others) to get a smart home into the news again – and it's still only experimental. The day when we (or at least everyone in Japan) will be living in such a house is still a very long way off.

Of course there will be much user resistance as with all new technologies – maybe people think such automated systems will make life too boring for them (like automatic transmissions as opposed to gear shifts on cars) and make them less likely to exercise their minds in trying to decide what to cook for dinner tonight or whether the bread will last until tomorrow. On the other hand, maybe their quality of life will be substantially increased because they no longer have to worry about mundane chores like remembering to get milk or turn the lights off or whatever. On still another hand, the timing has to be right – the technology or information has to be needed, there probably has to be more market pull than technology push.

Information comprises data and there is simply masses and masses of data around – collected and being collected, available and useable in some form or other. There are earth observation satellites circling the earth snapping its surface continually – there are terabytes and more of raw data which requires processing to turn into recognisable images – but only the tiniest fraction of the stuff collected is being actually presently used – and there is little hope of ever processing the remainder, but nevertheless it is still being collected. Some of the data, the images, are useful to only a small audience – for example, the fascinating photos sent back from the surface of Mars or Venus – but other remote sensing images are far more useful to governmental bodies and others for land management, transport planning, housing development, agriculture and so on. As a spin-off from this, imaging software which has the potential to stabilise video, remove flaws and eliminate blurring and thus reveal important details (i.e. information) which otherwise remain hidden is being developed by NASA. It is expected that such a system will be useful for medical imaging, scientific applications, home use as well as for helping to solve crimes.

If we worry about the data that our fridge might be collecting and passing on, about invasion of privacy, we have to remember that there are already all kinds of sensors everywhere gathering all kinds of details and images – and like the fridge, usually for benign, helpful purposes which try to anticipate your needs. Every time you access a web page, cookies are keeping track, every time you use your bank or credit or shopping card or fill in a form, more details are being gathered. Do a search with AltaVista on any topic and Amazon.com pops up with some books relating to what you asked. With interactive cable systems or personal computers then television programmes as well as web pages can also link up with suppliers and others not only to offer choices based on your profile, but also to promote and push even more personal services and products tailored to individual tastes. Electronic shopping, electronic banking, e-commerce – this kind of thing is the future of information; put another way, the future of information lies in the technology and applications which make the mass of collected data useful to the vast and omnipotent consumer market. And here is just a case in point: four volunteers were recently locked away for 100 hours to see if they could survive with just a computer and a credit card for £500. Using the Internet, they were expected to take care of all the basic human needs – buying, payment and delivery of food and clothing; creating their own entertainment; and seeking out contact and company via e-mail and chat. You can't call it really living, of course, but it is certainly indicative of what you can do with information and points to the way things are going!

References

1.   Raitt, D. I. 'Look – no paper! The library of tomorrow.' *The Electronic Library* 3 (4), October 1985, pp. 276-289.

2.   Anderla, G. J. 'Information technology 1970-2000.' In *Proceedings of 9th International Online Information Meeting, 1985 Dec 3-5, London, England*. Ed. D. Raitt. Learned Information, 1985. pp. 1-6.

3.   Lancaster, F. W. (ed) *Libraries and the Future: Essays on the Library of the Twenty-First Century*. Haworth Press, New York, 1993.

4.   Raitt, D. 'The library of the future.' In Lancaster, *op cit*, pp. 61-72.

6

Information Everywhere

Erik Davis

All of us carry crystal balls inside our skulls, so I can't claim to have any
special vantage point on the future of information. I can only report the
rather paradoxical message I came up with when I shook up the gray
Magic Eight Ball that rattles around inside my skull: in the twenty-first
century, the message went, information will become at once more invisible and more embodied.

Now, as anyone who has charted the last few decades of technology guru-
dom knows, brain-based auguries like this are often only marginally more
perceptive than the old rip-out-the-entrails kind. Nonetheless, I think
there's something in this vision, something that depends on our changing experience – both cultural and technological – about where
information resides. But we will get nowhere with such speculations until
we touch on a basic, but by no means simple, question: what is information? There are scientific descriptions of information, of course, but these
accounts hardly encompass what we generally mean by the term. Anyone labouring in the digital economy knows that information is always
in flux, but we tend to forget that the very concept of information itself – its
definitions and cultural reach – is constantly mutating.

In everyday terms, 'information' suggests a useful chunk of reified experience, a crisp and relatively manageable unit of intelligence lodged rather
low on the hierarchy of thought, somewhere between *data* and *report*. The
higher you go up this ladder, the more internalised things get – we think
of *understanding* and *insight*, not to mention the old fossil of *wisdom*, as
arising from within the human mind. Information is more tightly tied to
the external world, less a cognitive process than a kind of object or commodity that emerges from the spark gap between mind and matter – a bit
of sense that makes sense. Information tells us something specific about
the matter and energy that compose the ever-changing state of the world.
It's the scorecard to the game. As such, information also seems to enter
our minds from the external world, snugly bound to mundane materials
like newsprint or sound waves or web browsers.

We may pride ourselves on living in the information age, but information
has been a major player in human reality since the rise of the ancient city-

state. From the moment the first scribe took up a reed and scratched a database into the cool clay of Sumer, information has been an instrument of human power and control. But it was only in the twentieth century that information really became a thing in itself. People began to devote themselves more and more to collecting, analysing, encrypting, transmitting, selling, and using the stuff. Even more significantly, they built machines to automate and perform these tasks with a level of power and efficiency that grew far beyond the builders themselves, and this information combustion fueled the expanding apparatus of science, commerce, and communications.

In the middle of the century, scientifically rigorous definitions of information began to appear, definitions that were destined to invade biology, social science, and popular culture. In the late 1940s, a Bell Labs researcher named Claude Shannon announced the birth of information theory, an abstract technical analysis of messages and communication. The theoretical tools that Shannon created concerned any scenario in which a message is passed from a sender to a receiver along a communications channel – in principle, they can describe a conversation in a bar room, the replication of genetic material, or live broadcasts of the World Cup bounced off scores of satellites into hundreds of millions of homes across the planet. In order for the message to reach its goal, it must survive the onslaught of 'noise' – the chance fluctuations, interference, and transmission errors that inevitably degrade signals as they make their way through a distorted and analogue world. Shannon's celebrated second theorem proved that any message can be coded in such a way that it can be guaranteed to survive its journey through the valley of noise.

All this was great news for Shannon's employers, who were then multiplying telephone lines across the United States. But like the sciences of complexity and evolutionary psychology today, information theory also became a Big Idea, a conceptual model that people in many disciplines hoped would revise and clarify the known world. Once information received an abstract and universal form, it somehow became more *real* – not just a word or a squiggle on some Bell Labs blackboard, but a force in the world, a mathematically objective yet essentially mind-like material that could help explicate any number of seemingly unrelated phenomena by boiling them down to the crisp, binary unit of the bit.

So in the 1950s and '60s, social scientists, psychologists, biologists, corporate managers, and media organisations began reimagining and reorganizing their fields with 'information' in mind. The cybernetic paradigm of data-flow, with its nuts-and-bolts picture of signal and noise, sender and receiver, feedback and program, began to invade humanist discourses, promising to efficiently clean up all sorts of messy problems concerning communication, learning, thought, and social behavior.

Though the world has become far more liquid and far less 'top-down' since the days of IBM mainframes and Cold War institutions, we are the postmodern heirs to the technocratic conviction that the way to a society's heart is through its mechanisms of information exchange.

As the category of information continues to grow beyond the confines of communication engineers into the culture at large, it collides with the more amorphous category of *meaning*, that mysterious food of mind and soul. If you have ever attempted to really analyse toddler's talk or Eliot's 'The Waste Land', you'll know that meaning is a notoriously slippery beast to delineate, even in strictly linguistic terms. But one thing is clear: the redundant bits that information theorists deal with are only tangentially related to what most of us mean by meaning. From a technical standpoint, the latest chart-busting pop CD is full of information, but its relationship to meaning is far more questionable. And yet today many people confuse information and meaning, a situation which has put us into a rather unsettling bind: our society has come to place an enormous value on information even though information itself can tell us nothing about value.

In other words, the wide variety of values that help people judge the utility, validity and even morality of information have today been narrowed to a single value: the market. When traditional values butt heads with the information flows that now form the lifeblood of the economy, information tends to win – and not just for technological reasons. The apparently tremendous rewards of the digital economy are enough to placate any lingering concerns about the meaning of our technological plunge. And so more and more of us spend more and more of our hours producing, processing, and transmitting information, even as we wire our central nervous systems into a seemingly infinite matrix of data flows – corporate networks, news feeds, email, cell phones, PDAs, pagers. The Internet demands a constant and rapidly-changing influx of information, and everyone – grandmas, Wall Street bankers, techno musicians – have been captured by the conviction that those who navigate these turbulent waters are surfing towards a Byzantium of riches and personal fulfillment.

But the jury is out on whether the information age is making us knowledge workers or data slaves. Human beings are still creatures of meaning, with amazing but limited mammal brains, and it is not altogether clear how long we are going to stay afloat in the boiling seas of data. The networks that connect us can feel a lot like marionette strings, pushing and pulling our increasingly reactive, fragmented, and overwhelmed selves. Today, even the most trivial of tasks – selecting a new telephone, consuming the day's news, buying a bike – can spiral out of control, becoming dizzying, option-laden activities that demand a growing amount of our diminishing time and attention. Though we are convinced that we truly desire this cornucopia of choices, being an 'informed consumer' is now almost a full-time occupation. When you consider the information flows that data workers must surf in order not to drown, the situation becomes almost intolerable.

The tsunami of choices, facts, breaking news and new technologies is unlikely to abate, but I suspect that our sometimes pathological reactions to the stress of information overload will eventually help produce one of the two transformations mentioned above: the increasing invisibility of information. Think of the number of replies you get when you type a search term like 'used cars' or 'encryption' into Excite or HotBot. Without the filters and editors that define traditional media, online information is a glut of chaff with the occasional nugget of wheat. Of course, one man's husk is another man's kernel, and the ease of entry to the Net creates a more democratic media environment, at least in principle. But despite all the heterodox information one can dig out of the digital jungle, the Internet is nonetheless systematically overwhelming.

Human beings are protean creatures, and we will grow more accustomed to information excess, just as we now take the speed of trains, planes and automobiles for granted. This does not mean that data pathologies will abate, let alone disappear. Human bodies were not wired to sustain the level of stress that our information- and choice-saturated lifestyles generate. Nonetheless, kids born in the 1990s have already grown up in an environment defined by computer games, channel-zapping, portable consumer electronics, beepers, downloadable music files, virtual reality, and robot toys. They are going to adapt to data flows that drive older folks nuts, just as they can dance to music that even 1970s disco fanatics would find insanely fast. For the generation that comes of age in the third millennium, information will become more invisible simply because it will take on the characteristics of air – a plentiful, ever-present background of energy and potential.

This is primarily a cultural transformation, but there will be technological reasons for the increasing invisibility of information as well. Information environments, most certainly including the Internet, will give rise to sophisticated mechanisms that will automatically help search, weed, organize, and act on information. Of course, many technologies, e-publications and portal sites already filter information, both editorially and automatically, but these outfits will pale in comparison with the coming wave of 'intelligent' but non-human intermediaries: agents, bots, robust interactive interfaces, collaborative filters, smart newsreaders, 3D imaging techniques. At the same time, sites maintained by humans will continue to serve a crucial curatorial function, even as they replace the failed wave of universal portals with niche sites based on sensibility, lifestyle, religion, and politics. As more people turn to such entities to make decisions and keep up, the raw streams of information such sites depend on will become more invisible.

How useful these intermediaries will be is a whole other story. Any time you shift judgment and control away from individual people, you raise a host of questions about security, trust, and hidden bias –especially on the Internet, where the barrier between information and advertising is tissue-thin. The hidden political prejudices and absurd lacunae that characterise the current crop of automated net nannies are legion. But it seems inevitable that sophisticated and powerful programs will come to serve as gateways between human minds and an increasingly overwhelming information space. As voice recognition, speech synthesis, animation and AI improve, we may even come to think of these intermediaries as living characters – celebrity guides to the nets. Just as people turn to favorite magazines, web sites or radio personalities to get their news and opinions today, so might we make alliances with these autonomous avatars of information.

Other technological changes will lead to the apparent disappearance of information as well. Today, we remain focused on our desktop machines: portals into an information space that seems to be somehow 'inside' the box. But future computing will be distributed across an insanely wide range of small, portable, and hidden devices, and this distributed computing – some of it receptive to speech-driven commands – will erode the sense that information is crammed inside a machine. Given our seemingly unquenchable need to network technologies, these increasingly smart devices will be wired as well, which in the twenty-first century means they will be largely wireless. And wireless means that information slips out of the material networks of cables and routers and disappears into the ether. Like technological animists, we will soon find that the objects and spaces that surround us, both virtual and real, are laced with mysterious electronic intelligence.

The invisibility of information I am describing is primarily a cultural, even psychological process. One important historical analogue for this disappearance is electricity. With the dawn of the telegraph in 1844, the crackling fluid entered into the fabric of modern civilization, and by the end of the century, people were living in a new world of electric power, indoor lighting, and telephones. Electricity was the new wonder of the world – in Paris, ladies wore electric jewelry, while Washington D.C. congressmen 'took' electricity the way Tokyo businessmen snort canned oxygen today. But nowadays, electricity is, for most folks, a total bore. We have absorbed electricity into the technological unconscious of our civilization. The fact that basically the entire infrastructure of media, communications, and computing depends upon the electric grid and the electromagnetic spectrum is largely invisible – unless a blackout occurs, or industry deregulation creates new market opportunities. My intuition tells me that in another century, and in some way that's tough for our brains to wrap around today, information flows will similarly dissolve into the cultural background.

Of course, however deeply electrical power and its wonders are submerged within the technological unconscious, they remain dependent on material infrastructure. Absent this grid, the modern world as such could not exist. Similarly, the information flows that economies and communications require rest on material infrastructure, without which all the fruits of the new economy would wither on the fiber-optic vine. In this sense, we can think of electronic information itself as a kind of electric energy – a virtual fluid that, depending on the lay of the distribution land, flows and gathers in certain zones of the planet while barely touching others. We have created lush jungles of information, but also data deserts and frozen wastes of encrypted and proprietary records. And wherever you are, you need a plug. Without access – a gadget, an interface, and the education to use them – you are as helpless in the information age as a fellow trying to tune into a 50,000 watt radio tower with a dowsing rod.

It is important to stress the material basis of information, because the cultural story we have told ourselves about information so far has largely emphasized the incorporeal dimensions of the stuff. As the scholar N. Katherine Hayles notes in her book *How We Became Posthuman*, at some point in the postwar development of information theory and cybernetics, information lost its body. Though technologies like the Internet open up a new zones of mind, information does not emerge from some Platonic realm. The popular mythology of cyberspace is the most obvious example of this data-driven flight from matter – in fiction, the press, and newsgroups, cyberspace was described and imagined as an ethereal, almost otherworldly plane composed of information. Many of the early Internet proselytizers underscored the non-material aspect of the Net, arguing that cyberspace created an unbounded space free from the material constraints imposed by states and geography, as well as the social limitations that arise when bodies encounter bodies (gender, race, handicaps, even shyness).

As I suggest in my book *Techgnosis*, there is something almost religious about the cultural desire to vaporize matter and material processes into a stream of bits. The new economy alone cannot account for the fact that we are collectively porting everything possible onto the Web. At the same time, this 'informisation' of the world piggy-backs on the fact that money – another abstract, quasi-mathematical unit – has become the universal arbiter of value. Money is now the great translator, capable of abstracting and recoding the most obscure pockets of reality – DNA, the deep sea floor, even peace of mind – into commodities and units of exchange. Indeed, in an age when cash disappears into bits and fortunes are made off of millisecond fluctuations in the market's digital ganglia, information and money are now inextricably entwined. They are the abstract and virtual codes that undergird our global civilization.

But I hope, and to some extent believe, that the twenty-first century will also be a time when information becomes more embodied. Again, I mean this both in terms of technology and what we might call cultural mythology. According to the vision of distributed computing mentioned above, chip-based intelligence will spread throughout our manufactured world; it will also spread through nature, whose declining fortunes will be mapped, tracked, and monitored from the jetstream to the magma below. So while information becomes more invisible, it may also become more embodied, simply because its flows will literally penetrate the world. Information will no longer lurk in cyberspace, but will be woven throughout our commodities, dwellings, vehicles, even our bodies. Information will be part of the world.

Moreover, as computers improve our ability to model complex systems like the weather or the body's immune defenses, we will come to think of such systems as, in some more-than-metaphoric sense, dynamic plays of information. Finally, with the rise of genetic engineering and new reproductive technologies, the biological world, including human biology, will increasingly be seen as material expressions of genetic information.

Of course, as many social critics warn, the dominance and ubiquity of the information paradigm may make it even more difficult to restore the vital sense that we are embodied social actors sharing a finite and fragile environment. This danger is particularly pointed in the case of genetic engineering, which commodifies life itself, recoding the most basic birthrights of earthly creatures according to the abstract grids of information. Moreover, the omnipresence of surveillance systems, which are devoted to gathering information, raises profoundly disturbing questions regarding privacy and social control.

But what I'm calling the embodiment of information may also represent a potential turning point in our technological civilisation's basic relationship to the larger world. For one thing, it may help us expunge the dangerous and sloppy idea that the accumulation, consumption, and automation of information is, abstracted from material and social goods, beneficial in itself. That the information economy delivers profit is not a sufficient argument, because profit has also been dangerously abstracted from the larger context of life on earth, especially from the natural ecologies that form the ultimate matrix of human existence. As with economic gain, information must be seen as interdependent with the total world outside our windows – or our Windows.

It should come as no news to you that our excellent and ancient biosphere is seriously out of whack. One would hope that we might learn to direct the enormous productivity, novelty, and power of computers and information technology towards the restoration and establishment of an environmentally conscious civilization. Perhaps nanotechnology or some other visionary gadgetry will help save the day, but until such time, it's tempting to so that the next century will be green or not at all – or at least not worth living. Information and its market technologies have emerged as major engines of historical and evolutionary change, but from a broader, more holistic perspective, data and dollars are nothing more than energetic flows nested within a larger planetary system. And it's toward that larger system that we must now turn our attention, desire, and technological savvy.

7

Information's Golden Age: Substance With Style

David Skyrme

Some twenty years ago an IBM advertisement proclaimed, 'This was supposed to be the nuclear age; it has turned out to be the information age.' Around that time proponents of nuclear fusion felt it would not be too long before electricity would be too cheap to meter. With free Internet services on offer, it seems to have happened to information instead of electricity! However, even before we have fully grasped the potential of the information age, a new age is upon us, that of the knowledge economy. So what does this mean for the future of information? In this chapter three interrelated aspects are examined – information in its wider context, the challenges of managing information and the new opportunities for exploitation and trading. The chapter concludes with a fantasy (but highly feasible) view of knowledge zones based on lifestyles.

Pervasive and Perplexing

We need little reminding of the information glut. Statistics about the volume of printed information abound. There are ten times as many periodicals as just a decade or so ago. As much has been printed this century as in all earlier centuries together. The situation for online information is similar, with traffic on the Internet doubling roughly every 100 days. And if you think you are swamped today, you ain't seen nothing yet!

The portable computer into which I am typing these words holds the equivalent of half a million printed pages, that's 25 filing cabinets worth or a thousand books. Ten years ago, it was a hundredth of that. Look ahead ten years into the future and your hand-held computer will be able to hold more than the printed material in most town libraries. We will store information in many different places – in pocket computers, smart cards, cellular phones, even chips that you can implant under your skin.*

---

* See Kevin Warwick's chapter 'The Man With X-Ray Arms – And Other Skin-Ripping Yarns' in this book.

Information will be even more compact, portable and easily distributed than it is currently. Today the Internet connects some 200 million users. But this is only scratching the surface. Ten years from now several billions will be connected, as well as many machines, household appliances and scores of other devices. The best way to think of the future of information is that whatever you want, if it exists, you can probably get it, when and where you want it.

Yet despite this information glut, we still commonly hear the plea: 'drowning in data yet starved of information'. So do we face information feast or famine? Like food, we may have enough, but it is often in the wrong place at the wrong time. We suffer from inefficient retrieval mechanisms, poor memory, and may not even know what we already have. We are bombarded with information – some free, some expensive – but unless we are familiar with it, it can prove very perplexing. Proliferation and choice creates confusion and complexity.

Overload and Opportunity

Various surveys have shown a growing incidence of stress associated with being swamped with information. Yet cut off the supply and many managers will probably feel equally stressed about what important information they might be missing. The issue, of course, is that they need access to quality information. In my knowledge management work, one of the biggest problems that many working professionals face is the 'one million hits on AltaVista' syndrome. There is clearly an advantage, and a price premium, in having information filtered and organised according to your needs. That's the advantage of many online services and the facilities provided by a good library or knowledge centre. Similar technology to that which gives you one million 'hits' can apply relevance ranking, perform information filtering, and send out intelligent agents to roam the net to find new items of interest.

Users, despite their clamour for more precise information, actually also need knowledge. For example, a holiday brochure can tell you only so much. What the prospective holidaymaker really needs is knowledge that is often not in the brochures. They want other people's views and experiences. They need the kind of practical information that you find in publications like the Rough Guides and the Lonely Planet Guides. Simply bombard people with more and more information and, if it's not relevant at the time, it is not used. If you are an organised user, it might get filed somewhere for future use, and if you are even more organised it may actually get retrieved when it might prove useful. But this is the exception rather than the rule. It is no accident that some of the more popular web sites are structured portals, such as Yahoo!

A key opportunity for the future lies in well managed portals to knowledge. They are places on the Internet to access what you want, when you need it – delivered to the device that is convenient to you at that moment. But the good portals will offer much more. They will add capabilities that are seen in the best knowledge centres today – meta-descriptions of information – its applicability, its authoritativeness, the experiences of other users. Value will be added through multimedia clips – not the Encarta 30 second variety, but a whole half-hour tutorial if that is what you need, or a step-by-step video walk through a domestic plumbing problem you are trying to solve. You will also be able to dialogue with human experts as well for an additional fee. Users will get more discerning about what information they use, and whose knowledge they value.

Packaging and Pricing

The last few years have seen a surge in the growth of free information, whether trade magazines or Internet content, often much to the consternation of established suppliers charging over-inflated prices! Nevertheless, there will remain a strong market for paid-for information. After all, time is money. Therefore professional users of online sources are happy to pay for organised and customised content. The price reflects the time taken to analyse, organise, package information and make it easy to access. As the Internet develops, the business models in information markets are changing, just as the Internet is changing ways of doing business in other markets. A growing proportion of users will want freedom to compare and contrast information from different suppliers and pay-as-you go rather than paying steep up-front subscription charges. Contributors of information, particularly individual authors, will expect higher recompense than they typically get from publishers. After all, printing and distribution costs are a fraction of what they are for conventional publications.

Another anomaly in current markets is the vast difference in the price of information and knowledge. A day's consultancy will cost scores more than a good management book covering the same ground. What do you get for the difference? Certainly you get knowledge specific to your context, and may well get access to the specialist knowledge held at your consultant's knowledge centre. But this high differential means that consultancy models are likely to change. Given access to such consultancy resources, you may well prefer some ad hoc guidance as and when you need it, in short bursts (say a telephone conversation) rather than in more expensive daily consultations.

As a result of technology advances and the different perceptions of value, new ways of packaging, buying and selling information and knowledge are emerging. Already there are forerunners that give indications of how these new markets will develop:

- Online events, either synchronous (such as in webcasting) or asynchronous, such as in the Knowledge Ecology Fair (described in I3 UPDATE No. 16 at http://www.skyrme.com/updates/u16.htm).

- Information providers offering 'knowledge' on a pay-per-view basis such as Newspage (http://www.newspage.com) for industry and business news.

- Management Consultancies. Their business is knowledge, but they are increasingly packaging it, both for internal use (on their Intranets and Knowledge bases) and externally, such as Arthur Anderson's Global Best Practices (http://www.arthurandersen.com/gbp/) and Ernst & Young's ERNIE (http://ernie.ey.com). As well as access to databases, clients can ask questions, which are routed to the most appropriate experts for answer.

- Specialist 'portal' sites that act as gateways and links to many resources or help users through an individual process. For example, Career Mosaic (http://www.careermosaic.com) is a recruitment site that also offers hints on writing CVs, giving links to recruitment fairs etc.

- Problem solving brokers such as Teltech Resources of Minneapolis. It has a network of experts and a thesaurus of knowledge. As clients call in with problems, a knowledge analyst can help find experts who can solve their problem. Internet equivalents are medical help sites such as Doctor Global (http://www.doctorglobal.com).

- Electronic communities, many of which have a trading element, such as Geocities (http://www.geocities.com).

- Auction or brokerage sites that link buyer and seller, and allow online bidding. eBay (http://www.ebay.com) is the best known, offering over 2000 product categories, but there are specialist sites for niche markets such as cars (http://www.autobytel.com) and agriculture (http://www.agriculture.com)

- Markets in intangible products, such as financial futures, patent licences, copyrights etc. For example Alba allows integrated circuit designers to check intellectual property rights and licence and trade 'blocks of intellectual property' i.e. integrated circuit (IC) design elements to save time in new designs.

- Amazon.com – familiar to many as a way of buying books online (though delivered to your door!) – also has additional 'information' features, such as customer reviews, 'what other titles buyers of this book tended to order' and emails you with new titles that match your interests.

These examples illustrate a wide variety of information and knowledge types, from 'explicit' knowledge in small blocks to expertise in people's heads. It also shows variety in how users find what they need – online searches vs. human intermediaries. One of the widest differences is in pricing, how it is set, e.g. one off price or payment by use, and actual payment, off-line through conventional mechanisms, or online through an account or credit card. One of the anomalies of pricing is that the same information has different value to different people at different times. The dynamics of the Internet means that we can expect to see pricing set by users in reverse auctions where suppliers bid against user requirements. We can also expect to see dynamic pricing where the price is adjusted based on supply and demand. In summary, new technologies and entre-preneurial companies are creating a dynamic marketplace where information and knowledge is packaged into personalised portable pack-ages at the time and point of need.

Trading and Trust

The numerous permutations in packaging and pricing provide innova-tive opportunities to create new ways of information and knowledge trading. However, in the current embryonic state of Internet commerce, finding precisely the information you need and knowing that you are getting good value for money is rather a hit-and-miss affair. Online knowl-edge markets will only start delivering their potential when the following features are in place for every knowledge domain:

- Well-organised knowledge schemas – so you can quickly visualise the context and find what you need. Search engines can back it up, but these should allow search on fields such as date, owner, and subject.

- Good descriptions of knowledge content – to quickly qualify whether what is found is what you really want.

- A sampler (try-before-you-buy) or validation mechanism. This can be incremental pricing (starting with free samples) or some quality criteria by an accreditation authority (this is especially needed where the knowledge is more tacit), or credible testimonials.

- A fair and transparent pricing mechanism. At the moment you can find some items on the net that cost tens of dollars at one URL and are free at another! Perhaps there is a role here for consumer meta-guides.

- Simple and easy payment, including micro-payments for small value items (e.g. for the $1 sampler). The notion of a penny for your thoughts is closer than you think – a person with an embedded chip agrees to answer a colleague's question, through Internet phone, for example, and while he or she is speaking his embedded electronic wallet receives a payment transfer.

- Mechanisms for tacit knowledge exchange between buyers and sellers and amongst groups of buyers. This could be through web conferencing, but also parallel Internet channels (e.g. one giving voice communications with a sales representative and another a shared view of their Internet pages).

- Price mechanisms that are dynamic and context situation sensitive according to users' needs (something like the situation today where two airline passengers sitting next to each other have paid widely different prices, though unlike aircraft you can sell the same information again in future!).

- Use of intelligent agents as online traders, for example dispatching an agent to find you the cheapest information on a particular subject.

This last approach is likely to become common once there are agreed standards and structures (using XML) for information 'wrappers' that describe what content is inside. Another cautionary note: we may need some regulation in these knowledge markets, for example to prevent intelligent agents playing havoc with market stability. Experiments have shown that agents negotiating with each other can cause wild pricing fluctuations as can happen in computerised stock exchange trading.

With mechanisms such as those listed above, knowledge flows will improve, the buyer community will be more knowledgeable and the seller gets feedback on products and develops closer relationships with their customers and buying communities. An idea of how some of these features are already being considered is through the IQPORT trading platform (http://www.iqport.com) and its guilds such as BRIGHT (http://www.bright-future.com). Incidentally, in this market the author gets close to 50 per cent of the value of the transaction – much better than conventional publishing.

As online trading evolves, one element that is really important is that of trust. How can you trust that your money is well spent? How can you trust that the buyer or seller is reputable? How can you validate quality, reliability, currency (i.e. that it is up to date) etc.? Sometimes it seems that even well known brand names cannot be trusted. My trust in the *Financial Times* was destroyed in believing an article that announced that Guinness had paid a large sum for the rights to call GMT (Greenwich Mean Time) 'Guinness Mean Time' during the millennium year. It was a good knowledge-age example of the value of an intangible. Unfortunately it was an April Fool PR stunt by Guinness, even though it was published before 1st April. Now I am wary and double-check anything that sounds out of the ordinary, which – since the rules change so fast – is almost everything. A year of so ago, free Internet accounts might have seemed like an April Fool's joke, but over one million users of Freeserve (http://www.freeserve.com) suggest that it is not!

What we can expect to see is the emergence of trusted third parties (who don't play April Fool's jokes!). These are already planned for e-commerce trading, where a third party will look after encryption keys that allow both buyer and seller to transact business with security and confidence. This approach can be adapted to information, where there can be information accreditors, who can validate, as far as is reasonable, information from an unknown provider e.g. through peer review or reputation. Just as the credit card industry has information providers who provide details about an individual's financial dealings and standing, a similar service could build up around the information industry, provided of course, users are prepared to pay for these services in preference to their own judgement.

From Function to Fantasy

As we enter the next century, the basic material needs for most people in the developed world are largely met – even though many crave for more. Our individual lifestyles will be determined less by the need to acquire basic necessities but more on our individual values and desires. More people will have more time to seek personal fulfilment in a wide range of work, family and leisure activities. Information and knowledge will enrich most of these. Just as in the industrial era, when production capacity is more than sufficient, there will be growing choice in the way that information and knowledge is packaged and consumed. A golden age of knowledge draws us into the new millennium as we master these capabilities. Information and knowledge will surround every activity we do. For purposeful work, relevant knowledge will help us achieve our aims effectively and to the best of our ability. In our leisure activities it will help us gain maximum enjoyment. But this will only happen if the information and knowledge is tailored to individual needs. As consumers, each of us has preferred ways in which we like to consume.

Thus we can expect to see the emergence of different knowledge clusters and access methods according to individual life style and consumption preferences. We already see themed areas in cyberspace. There are some places I wouldn't dream of visiting but that my daughter relishes. The way that information is provided will also reflect your current mood. Perhaps today you are in serious mode, but later you may in relaxed or leisure mood. Therefore information will be presented in many exciting ways – multimedia, multi-channeled, even in virtual reality – and that goes for business information too.

As a 21st century knowledge navigator, you will set your mood button on a personal knowledge transporter – or simply speak to it. It will work through your activity list, and bring the information you need into your immediate environment. It may be displayed on large flat screen on your wall, as holograms in front of you, or simply work seamlessly in the background on a range of computerised devices. A quick push of a button or speech or eye movement command and your transporter converts to a person-to-person communicator or a hand-held personal assistant. Just as different retailers have different formats, your physical surroundings will offer a range of different environments. You will choose your place for work or leisure (when not at home) based on the kinds of environment you prefer and the kind of company you like to keep. Some may be whole communities rather than specific locations in a general locality. Others will be entirely virtual. Each knowledge zone will develop its own ethos and set of values. These will be closely associated with similar zones around the world, so that as you travel you can always find a comparable zone should you want to – just as hotel chains provide a common look and feel wherever you are. Sometimes you might want a straightforward functional zone. At other times you might want a fantasy island. The choice is yours.

And when you need a break from this intensive information immersion, you will seek out an information free zone – but then there might be no information to help you find your way out!

8

The Future of Information Access

Colin Steele

The number of users of the Internet is estimated to reach over 300 million by the end of the year 2000. As high bandwidth access becomes the norm, the convergence of media available on the Internet will become ubiquitous. Text, audio and video will be streamed. Access will be changed in profound ways by the presence of software that transforms the one-way media into interactive resources. As Alberto Manguel has written in his seminal work *The History of Reading* (1997), the cyber generation returns from the book-centered Hebrew traditions of Augustine to the bookless Greek tradition. Virtual reality replaces textual reality. The future is difficult to predict but certain trends seem inexorable. Globalisation, aggregation of providers, the development of niche markets and a bewildering array of information sources on the one hand and commercial homogenisation (Americanisation?) on the other. As David Brin wrote as far back as 1990 in his novel *Earth*, 'And to think, some idiots predicted that we'd someday found our economy on information. That we'd base money on it! On information? The problem isn't scarcity. There's too damned much of it. The problem usually wasn't getting access to information. It was to stave off drowning in it. People bought personalized filter programs to skim a few droplets from that sea and keep the rest out.'

In the future users will by and large 'bypass' libraries to gain information via the web. It is argued that customers of such portals as America Online prefer the familiarity and user-friendliness of the AOL community than to go searching outside those gateways. People will rely on their friendly neighbourhood portal just as they relied on their friendly neighbourhood library. Libraries will still provide a physical social forum, for example for senior citizens, just as universities in a virtual university environment will need to provide a social environment for undergraduate students. Most information, however, will be garnered at the desktop facility. Access will be continually fostered through the need for lifelong learning and the skill-base training requirements.

Access will, however, be conditional on an ability to pay. The buying of goods over the Internet via encryption, smart cards, etc. is going to explode in the twenty first century via one-stop cheap digital boxes for TV and Net access and mobile web telephony. The music industry provides yet another example in terms of digital downloading. If one wants access to a major new release this will obviously cost, but if a new group wishes to put their music up on the Net for free, this will be a form of advertisement.

It would be fair to say that electronic licensing in the academic arena has generally led to more restrictive practices than was the case in the print environment. At the moment some major American and European publishers only allow access to material from a terminal in a library, while others have very restrictive site licences and will not accept multi-campus licensing for the same university. Other publishers aggregate their material and only offer discounts for that aggregation when much material in these information aggregations is not required by an individual university. Electronic access to commercial academic information currently is anarchic and is not helped by some multinational publishers being understandably more interested in profit returns than the effective distribution of scholarly information on a globally equitable basis.

One of the issues is to separate the quacks from the gurus in the information environment. Libraries can provide an important accreditation service, in the widest sense, by filtering information and approving information or providing multiply gateways to information. Recent surveys have shown that users want one-stop shopping to electronic full text, to do their own searching on the Net via such facilities and twenty-four hour access. Where an intermediary is involved it is for the purpose of a value-added service.

Virtual medicine or Internet medical web pages are now troubling some doctors as users can get expert information on the Net. Why wait for ages in a doctor's surgery and then pay a lot for a basic diagnosis if the Net doctor can help you? British Telecom's prototype of a cardiac monitoring wristwatch is just one example of medical technology to come.

Local, indeed national, providers, will have to compete with the providers in the global marketplace. In that process globalisation is going to be a key factor and the need to establish niche markets is essential. It is quite clear in a wired country like Australia, and one which is separated from a physical access to much of the world's information, that firms like Amazon.com have made and will continue to make massive inroads. It is certainly quicker to obtain material required urgently from Amazon.com than from local Australian booksellers who have to import the material by traditional methods, apart from the fact that the cost is cheaper from Amazon.com. This will inexorably lead to a decline in local employment and service unless niche markets can be established.

The same process could be said to be happening in the music and shopping areas in general and the questions of local revenues via taxes is important in an e-commerce environment. If Amazon.com can do this, albeit currently with major losses, in a bookselling environment one could see many of the principles in a generic sense being applied to the library and information sector. Twenty-four hour global reference desks, accredited information portals probably subsidised by commercials, use of multimedia digital course packs produced internationally, the rise of virtual universities etc. The net result, in more ways than one, might resemble an international global amphitheatre with market shares being debated as ruthlessly as the interaction between the lions and the Christians in the Roman amphitheatres.

If libraries do not add value to the process between creator and consumer, then they will disappear just as quickly as the horseless carriage. Libraries will more and more become print archives and will be utilised just as manuscript collections are currently visited. Many collections of information will not be digitised in the near future, particularly collections from Third World countries. The role of libraries will be one of repositories on the one hand and active filtering in the Net environment on the other. Physical environments such as Eastern Michigan State's new 'cybrary' and the New York Public Science and Technology Library Internet area are attractive venues which currently merge both facets.

Recreational reading has been excluded from this brief overview. The paperback book will have some considerable longevity but the future will also include portable hand-held electronic devices which in many ways will physically resemble a book. E-books will come in as many varieties and formats as print. The means of utilising them may, however, be different. Future generations will reflect in Net and 'reading' use their current use of computer games and multimedia habits of accessing information.

Don Tapscott, who used the term 'paradigm shift' to describe the impact of the information society, has indicated in *The Rise of the Net Generation – Growing Up Digital* (1997) that as the current generation has grown up 'bolted in bits', they will not only be familiar with the Net environment of non-sequential access, but they will also demand highly customised products. Children coming through as the Internet generation have lesser attention spans for print material and jump across information flows in a web/URL environment. Students now use basic search engines to look for the information on topics such as human rights before going to library catalogues or, dare one say, the shelves of libraries.

A crucial issue in the organisation of knowledge is to provide a filter to protect, particularly the young. This is however far from an easy issue to untangle. At the time of writing, the legislation passed by the Australian Liberal Government, ignoring high-level technical advice on the difficulties of blocking such material by the CSIRO (Commonwealth Scientific and Industrial Research Organisation) is causing concern by legitimate online providers. A study by Electronic Frontiers Australia indicated that one Internet filter based on text would exclude the entire Ozemail domain and such diverse organisations in Australia as the National Party site and those of the Christian Bookselling Association and St Lukes Lutheran Church, Nambour.

Many providers could be driven offshore and any government attempt to police access to overseas sites would resemble previous attempts at censorship as carried out by such places as China and Singapore. Internet information being so ubiquitous provides numerous opportunities and challenges. Responsibilities need to be devolved to the personal level but the fluidity of the Net makes censorship extremely difficult. No one wants free unrestrained access to paedophilia or bomb-making sites. Any enlightened society must attempt to crack down as much as possible but the total bludgeoning of a system could become even more counterproductive and produce an Orwellian situation of newspeak which narrows the range of thought so that no individual can publish subversive thoughts against the regime of the day (c.f. China and their blocking of the BBC web sites).

Neil Stephenson, in his essay 'In the Beginning was the Command Line' (http://www.cryptonomicon.com), has extrapolated that a computerised, visual interface culture linked by GUIs (Graphical User Interfaces) could lead to a world of desolate culture and the fact that people become remote from the physical world – there are many images of this virtual world in science fiction novels and movies. The societal impact of individuals telecommuting from home, or factories full of people at terminals, have also been depicted vividly in those genres. E. M. Forster's *The Machine Stops* is a classic historical short novel on the reliance on technology by the individual and the impact on them when society collapses.

Margaret Wertheim, in her latest book, *The Pearly Gates of Cyberspace* (1998), has indicated that cyberspace fills the spiritual vacuum in people's lives left by the decline of traditional Christianity. This may be taking it too far at present but the nature of human identity and the evolution of artificial intelligences is a long-term debate. This recalls an early 1950s short story by Arthur C. Clarke. The world's largest supercomputer was created. The protagonist asked the first question 'Is there a God?' and the answer was, 'There is now'. Information access has the opportunity to provide both the liberation of the individual and his or her entrapment. The major problem is that most of our administrators and politicians are unable to recognise, partly through a generational issue and the fact that the e-commerce drives their thinking, the major issues need to be debated in every society. The twenty-first century will be the information century.

9

Marketing in the Digital Economy

Don Tapscott

Marketing is undergoing its biggest transformation ever, paralleling the shift in many aspects of business and society from broadcast to interactive:

- The old broadcast model of *communications media* (print, radio, television) is giving way to a new interactive model. Rather than one-to-many, centrally-controlled, unidirectional, the new model is one-to-one, controlled by no one and interactive.

- Broadcast *learning* is being replaced in the schools. The one-size-fits-all, lecture-oriented, teacher-focused approach to instruction is being replaced by customised, interactive, student-focused learning which exploits networks and the new media.

- The broadcast model of the *firm*, where senior executives sitting at the top of rigid hierarchies, issuing decrees to compliant employees, has given way to more collaborative, two way, networked models.

- Even the old broadcast model of *politics and governance* is changing. Old model: 'I'm a politician. Listen to this 30 second video clip and then go out and vote. I'll broadcast to you for four years. Then we get to do it all again.' This is giving way to new interactive approaches to citizen involvement, including conversational democracy, the electronic town hall, online citizen brainstorming and more.

This shift is beginning to affect marketing, obviating most of what we currently think. Prices used to be established by sellers and broadcast to buyers. Brands used to be images established through one-way print and broadcast communications. Advertising used to be one-to-many. All this is about to end – buyers will establish prices. Brands will become relationships. Advertising as we've known it is through.

It is not just a change in technology causing this upheaval. Rather there is an intersection between a technology revolution and a demographic revolution – the rise of the Net Generation.

Today's computer-literate web-surfing kids are going to force corporations to rethink their marketing strategies. These kids are the first generation to come of age surrounded by digital information technologies. The effects are already dramatic. Compared to their parents, today's youth are more curious, self-reliant, contrarian, high in self-esteem, and global in orientation. And as consumers they will be much more demanding and discerning.

The 88 million offspring of the North American baby boomers – whom I have dubbed the Net Generation – now outnumber their parents by a healthy margin. They are the richest young generation ever, and unlike any kids before them, they also influence a large and growing portion of their parents' purchases. The Alliance for Converging Technologies estimates that American pre-teens and teens spend directly $130 billion and influence the spending of upwards of $500 billion.

It isn't only computers, video games and high tech purchases that these children influence. They feel they should have a large say in every-day grocery and clothing purchases and they expect to be consulted about major household acquisitions like cars and appliances.

In the past kids had little basis on which to exert such influence, but with digital media, particularly the World Wide Web, they now have a way to become more knowledgeable about a product than their parents. Consider the prospect of a thirteen-year-old girl influencing her parents to buy a Volvo based on safety statistics she has downloaded from the web.

But successful marketing to the Net Generation requires more than just packing the home page with buckets of product info. Web surfing encourages certain personality traits, and goods and services targeted at kids should reflect this. For example, Net Generation kids love options. On the web they can surf the world, flitting from page to page, idea to idea, always finding something new. Surfers don't like being boxed in with artificial constraints. If you want their loyalty, you must give them choice.

Moreover, they won't accept being penalised for the wrong choice. If they don't like a new web site, they just hit the back button. The same should be true of products and services. Mistakes should be undoable.

Net Generation consumers also want the ability to customise. Corporations should follow the lead of software companies that allows users to fine-tune their software to their own work habits. They also like to try before they buy, as the concept of the demo is deep in their online culture.

Whither the brand?

As these kids grow older, the market will increasingly become a collection of inveterate comparison shoppers, well-informed, opinionated, and demanding. In such a marketplace maintaining the concept of 'the brand' — at least as we've known it — becomes a much more difficult task.

Until now the brand was largely a product of mass communications. Using the one-way broadcast and print media, marketers could convince people through relentless exhortations to 'Just do it!' If you say 'Things go better with Coke' enough times you can establish the Coke brand in the market.

To date this technique has worked well, but it won't be so successful in the future. The Net Generation is building a culture that is incompatible with the mass communications necessary for current brand establishment. In marketing, interactivity equals increased power to the consumer to make informed choices and to buy products that deliver real benefits and value and reject those which do not.

The Energizer bunny can claim to outlast Duracell, but if it doesn't, media-savvy youthful shoppers will find out. They will go on the Net to examine third party evaluations or participate in discussion groups to determine which battery makes the Walkman work longer. Such forums have enormous potential clout. Witness Intel being brought to its knees over the buggy Pentium chip because of the furore raised on the web.

Smart software on the horizon – software agents – will further undermine the traditional branding. Rather than trusting the brand, kids will trust their agents. Sometimes called softbots, knowbots, or just 'bots' – agents are software which gets to know them, their preferences, and their sense of style. These tireless little workers surf the Net day and night looking for information you've requested: finding that perfect chocolate chip cookie; evaluating new movies based on your preferences and the opinions of others you trust; organizing your personalized daily newspaper; communicating for you; trying on different types of jeans and doing other jobs. In many areas, trusting your agent will become synonymous with trusting your own experience.

From Brand Image to Brand Relationships

Developments such as these will cause a change in thinking among marketers – away from focusing on brands and brand image to thinking about relationships with customers. The Net provides new opportunities to evidence the true value of products and services as well as to create meaningful relationships between providers and customers based on trust. As the power of mass communications declines, replaced by the power of the interactive media and therefore the consumer, brand loyalties for informed and value conscious purchasers will be based increasingly on value.

It is networked information that is bringing value and real benefits to the fore. For example, grocery shoppers using the Peapod web site can ask for all the products in a certain category sorted by different criteria such as calorie count or nutritional value. The most frequently used sort criteria are cost followed by fat content. Determining the healthiest peanut butter takes seconds, and Skippy's mass marketing would have little impact on the purchasing decision. Good brands will correspond closer to good products, and those products which are undifferentiated in value quickly become a commodity.

This means that companies need to focus even more on innovation, as value comes to the fore in an internetworked world. Tide can say it 'washes whiter' till the cows come home, but if it doesn't these media savvy consumers will find out – in 30 seconds. There will be agents and soft-bots and know-bots and third-party rating systems and discussions groups where it will be very clear who washes whiter. So P&G needs to focus even more strongly on innovation and having the best product.

It may not be the molecular structure of Tide that washes whiter. It may be the services bundled in with Tide and delivered through the Net and the various information appliance in the home – including the washing machine, the soap box, and the white shirt – all of which will have chips in them and communicate with each other. P&G changes from being a soap company to becoming a relationship company – helping the homemaker keep shirts white, clothes undamaged, the home cleaner and perhaps even deciding which shampoo to use on a given day.

Or take retail banking. There has been much discussion about relationship banking – but this is not about getting all your statements on one printout. Take the scenario of a 22-year-old graduate from engineering school. He gets a new job, new car and nice apartment. Then he misses a car payment. In an old-style bank, the car loan application generates a nasty letter suggesting he pay up. A digital economy bank has an internal IT architecture that can view him as a customer of the bank as a whole – a good VISA record, a retirement fund, parents' collateral on another loan, strong references from his employer and the like. Rather than harassing him, the bank reaches out to him with the Net – delivering financial planning services, helping him reorganise his finances and in doing so setting up the basis for a life long relationship. In this scenario the 'brand' is not an image established through the broadcast mode, it is a relationship, based on value and mediated through the Net.

The Transformation of Pricing

We're all pretty used to sellers establishing the price. There is the 'Manufacture's Suggested List Price,' defined by the makers of goods and the 'sales price' used by retailers. We learn about 'bargains' when retailers discount prices. However, the notion of a seller establishing the price is a relatively recent phenomenon, which is about to be turned on its head in a big way.

In ancient history there was no 'price,' since money didn't exist. Goods were exchanged by barter. Money didn't happen until 600 BC, when the Lydians hit upon the idea of making coins by stamping bean-shaped lumps of metal with official symbols. The idea quickly spread to all the important trading cities, with products now being 'sold' for currency. But the prices were established through auctions, such as the slave auctions of the old Roman empire, or through haggling as in the Athenian market-place or the Roman agora. The idea of a take-it-or-leave-it price being fixed by the seller really only became mainstream in the past couple of centuries. And with the Internet's arrival, it may quickly become a thing of the past.

We are seeing a whole host of new Internet-enabled pricing models. On eBay.com, millions of products, from Beanie Babies to luxury cars, have been sold to buyers around the world in auctions that last from three to seven days. The concept began with used goods, much like a global garage sale, but has grown to include new products.

On Priceline.com, it's the buyer that sets the price, saying what they are willing to pay for products such as airline tickets and hotels. Only if an airline or hotel is willing to meet the price will a transaction occur. Priceline.com has even extended the concept to cars. Buyers specify the car, the options and the price; Priceline.com will find out if any dealer in the area is willing to agree to the proposal.

The Internet gives the buyer much more knowledge of the market, it emboldens the buyer with confidence and therefore power. This new arrangement is ideally suited to the Net generation, which thrives on interaction.

Transformation of Advertising

The Net enables marketers to target individuals with messages that they will value. The best people to target with Volvo advertising are people who are looking for a car and who are in the Volvo demographic. The Net allows delivery of broader and more detailed messages to these people. The narrow message – 'Volvo is the safety car' – is targeted to a precise demographic group – yuppies with families. The messages are delivered through the mass media to the correct audiences (upscale publications, specific radio and television programs). Alternatively, the message is delivered through direct marketing, aimed at selected city boroughs, streets and even individuals.

But when the media becomes based on choice, messages, ironically, can become comprehensive. Volvo can become the 'everything car.' The youthful driver considers buying a new car. One day in cyberspace, she likes the look of Simon Templer's Volvo in the movie *The Saint*. She stops the movie and asks Simon to tell her about the car. She asks about Volvo's safety and learns that they have great safety features. She asks about acceleration and is told they do well there. She inquires about mileage and Volvos are competitive in this area as well. Says Dave Carlick of Poppe Tyson advertising, 'Instead of having one feature which suits many people, you can have many features which suit one person. Interactivity allows that one person to explore products and services according to their own interests and find out what is important to them.'

The evidence suggests that the N-Gen is on the cusp of this kind of proactive product exploration. Advertisers should create compelling interactive environments where consumers will want to go and can do comparative shopping and thereby receive appropriate and comprehensive marketing messages.

The Net will also allow ads to be integrated with each other and with transactions, in ways not possible in the physical media. Software company Oracle's President Ray Lane describes the situation where someone buys a Sony stereo online, and is offered a $10 coupon to the electronic Tower Records for any Sony title. The coupon expires in 15 minutes. He says 'that's a highly qualified, motivated audience of one.'

So far, marketers haven't progressed very far in moving to true interactive advertising. We've taken the old model (the 'banner ad from newspaper or broadcast media) and applied it to the Net. We need to take the next steps towards truly interactive marketing.

10

Assessing the Impact of Information in the Digital Age

Dan Wagner

The Future of Information – a phrase that will no doubt mean a million different things to a million different people. To me, it brings to mind a sense of opportunity, a sense of boundaries still to be broken, a sense of 'norms' to be challenged.

We have come far in the 20th Century – further than mankind could have dreamt at the start of the 1900s. We now have a medium in which information can be published to a much wider audience, and much more quickly, than the printing press could ever have achieved.

Indeed, we have come a long way from the advent of Dialog as the world's first commercially available online information service in 1972. Users are now widely taking the option of web-based interfaces instead of the Windows-based graphical interfaces that were a breakthrough only five or six years ago. Yet the command-based interfaces that have been with us for decades are still used and are still perceived by some users to offer additional speed and precision.

In what we call the online information industry, we have never had such a large selection of high-value information providers publishing electronically. We have never had such scope to reach audiences from the corporate library, to each desktop across the corporate intranet, through to partners and suppliers via extranets. The opportunity for information providers, information aggregators, information seekers, has never been greater.

So where do we go from here? Some commentators believe that we will all be wired into our computers by the middle of the next century (if not before then). Perhaps we will, but for the purposes of this contribution, I am going to look at the less invasive implications of the information age in the medium term, including:

- Further changes to distribution models
- Further changes to business models
- A re-think of what 'information' is available and how it can be used in the digital environment

Addressing 'Truths'

To misquote one of our most famous novelists, it is a truth universally acknowledged that everyone wants to publish online.

In terms of distribution models, an increasing number of publishers are reaching their audiences directly, over the Internet, as well as by print. Many of us have assumed that the low cost of Internet publishing will ensure that every company able to publish online will do so. Yet the explosion of information available over the Internet and other media is being seen by some as a factor which is moving them into more niche activities and – in contrast – away from publishing. For example, a research firm with which I am familiar has decided that it is not profitable for the company to publish generic market research reports. Instead, the company sees its opportunity in conducting pure custom research projects for particular clients. It is building its business model on that basis.

This of course is one of the great unanswered questions of the Internet age: which business models – both historical and new – will actually survive? There is more information available than ever before, but in certain areas business people are working in circumstances approaching an information vacuum. The Digital Age is bringing with it new ways of doing business, requiring new information.

The information upon which banks and backers traditionally based their decisions to fund new start-ups (and indeed continue to support those companies) has always been based upon as much 'known' data as possible – data such as what other similar businesses have earned, how they have been valued, when they were able to break-even and start paying dividends to shareholders. Yet the start-ups of the 'Digital Age' do not tend to operate to these kinds of tried and tested norms. The information simply is not available – the benchmarks in terms of valuations, break-even points, critical mass – are still being formed. In addition, those existing business models – which are seen to offer 'accepted knowledge' – are frequently being challenged by new entrants.

At the consumer level, for example, companies such as Online Originals are taking book publishing direct to the web, potentially creating a new business model in the publishing world whereby only the most popular titles will ever be printed. Yet this is, in effect, where that market is going anyway. Given the costs of printing and distribution, publishers are less likely than ever to take a chance on an unknown author. Many of the large publishing houses now prefer to concentrate on tried and tested bestsellers, whose performance is deemed to be more predictable, or those newcomers who are surrounded by some degree of hype that will guarantee sales.

The benefit of companies like Online Originals is that lesser known writers will also be published, although perhaps never in 'traditional' media.

The growth of these companies is assured not just because of a wealth of potential material, but also because the arguments of portability raised against such ventures no longer hold true. Even today, books are being published digitally in such as way as to be downloadable onto personal digital organisers, so will consumers wish to carry a book when a palmtop fits better into a pocket?

The Human Factor

There is, as I mentioned previously, a school of thought which believes that computers will govern us. After all, more data can be held in a computer than in the human brain and as an illustration of that, a computer has already beaten the best chess player in the world.

Chess is a structured game. There are definable moves, a clear objective. That is the sort of processing in which we know technology can prove most effective. Yet the world in which we live is still governed by the inconsistencies and subjectivity of human nature, and indeed the surprises provided by nature itself. Life is not wholly predictable, and cannot yet be truly understood by machines. The information available to each of us is not only what is said or written – what we see, hear or read – it is also what we feel or smell or can touch.

There are nuances in conversation between two people (or indeed a group of people) which go much further than what is actually said. Technology may be able to record and transcribe the spoken or written word, but it does not yet recognise the other information available to the human eye and ear. A tone of voice or a nod of agreement which indicates that someone is supportive of one course of action, or a shake of the head indicating that someone else is not. This, together with what a businessperson may call instinct, is the 'soft' information, so important in situations such as negotiation, that individuals must still recognise and decipher for themselves.

A Wider View of Digital Information

Given the way in which technology is developing, we are also able to perceive digital information as being more all-encompassing than ever before. Online information was once merely the result of the digitisation of existing text or information. There will increasingly be new types of information available, new ways of using information and the ability to create new information within organisations.

We have seen the way in which multimedia is increasingly used at home and in business. We are also seeing an increase in the use of technologies such as virtual reality, imaging and the digital control of environments such as the home.

One of the truly innovative applications of virtual reality is in its use to show schoolchildren what it was really like to live in the past. Rather than reading about life as it was, they are able to see it in front of their eyes. It is only a matter of time until this can be taken further, to incorporate the smells and tastes that offer equally relevant information about our environment.

Similarly, how history is recorded will also change. Projects being run in the US are encouraging children in disadvantaged neighbourhoods to come to multimedia centres, where they can see their own paintings and words digitised and put onto the Internet. In effect, they are creating their own mini-movies, records of their lives and thoughts. While this has the benefit of encouraging these children to become part of the information age, it is also giving us a wealth of information about life in our time, which might never otherwise be recorded.

Twenty, even ten years ago, children were encouraged to bury time capsules, to give future generations a glimpse of the past. The reality is that in the future, our children and grandchildren will simply be able to 'download history'.

We also know that the impact of the information age will soon be making its presence felt in the home. Many people in the UK already use loyalty cards. In addition, a percentage of UK consumers have started to purchase their grocery supplies over the Internet, from these same supermarkets. In e-commerce terms, the data gained from the loyalty card is invaluable, and can be used in conjunction with the 'Favourites' functionality offered by the service itself. A supermarket can analyse how frequently a customer has purchased milk or pasta or frozen goods in the past, and use that information to send the customer reminders over its e-commerce system. At particular times of year – say Christmas – it could remind the consumer that their shopping list the previous year had included certain meats, condiments and drinks, either offering the same items again, or related items.

Kitchen appliance manufacturers are also looking at integrating technology into refrigerators, ovens and microwaves. Your refrigerator could soon be reading bar codes so it can remind you when food is out of date, and your oven or microwave reading the same codes to ensure that pre-prepared foods are cooked at the right temperature and for the right amount of time.

So we will be dealing with more and more digital information in our everyday lives, ostensibly to encourage learning, to make life easier and purchasing more convenient. In reality, we will also be divulging more information about ourselves than we could have ever predicted.

Creating Information in Organisations

From an organisational standpoint, companies are now able to generate entirely new information to support strategic decision-making. By linking documents, spreadsheets, data from accounting systems and external research, an organisation is given the opportunity to create information that is specific to its own needs and circumstances. This will be facilitated by categorisation and search technologies, and by the widespread adoption of XML (eXtensible Markup Language).

Think about how information could be linked – if only the structure, technology and culture supported it. Take, for example, a document written to propose entering a new geographic market. Information held on the accounting system may reflect the potential value of that operation, and the probable cost of running it. Link that together with supply chain or inventory information – facilitated by e-commerce – and externally produced market research on the growth opportunities of that geographic region – and a more holistic picture may be constructed that assists timely and effective decision making.

In the hyper-competitive markets of the future, traditional ways of doing business will be too time-consuming and inefficient. Organisations will look not just to information systems which keep everyone in the organisation informed, but increasingly to technologies that will keep the organisation and its employees ahead of the game, by maximising the knowledge – both internal and external – at the organisation's disposal. Effectively, we are talking about taking the organisation's knowledge management and enterprise resource planning systems and merging the two.

If we once again take the example of the organisation referred to earlier, much of its information systems are already digitised and could be linked. If the organisation implemented an 'agent' to sit behind each of those digital information systems, creating permanent links between them, it would be possible to automate information searches and basic analysis of the information found, thus shortening the research cycle of many projects, and even suggesting new opportunities. The additional benefit, of course, is that the time of business development managers, researchers and corporate librarians could then be better utilised in more in-depth research and analysis.

The Future of Information will be about people as much as technologies, and making the most of both.

11

Entering the Mainstream: Digital Information Feeds the Business Ecosystem

Stephen E. Arnold

Metaphors in common use about information can be very misleading – not because they promise too much but because they say so little. The information 'environment', 'infosphere', even the 'information highway' do not do justice to the richness and the potential inherent in the digital tools that are now part and parcel of our daily lives.

The trajectory of change is clear. Information – particularly the digitised, network accessible, Internet version – has moved from the margins of professional life to the centre of professional activity. The same may be said of the impact of electronic network access on the personal life of upwards of 200 million people (and millions more each week). Furthermore, age is no barrier to the use and enjoyment of electronic information in its myriad forms. Children of tender ages play games with studied concentration. College students think nothing of sending parents and grandparents an e-mail 'to keep in touch', and senior citizens are embracing the online life with gusto.

In 1981, a Texas Instrument Silent 700 dumb terminal allowed a person to connect to a remote computer to carry out such tasks as querying a database. The process required a dial-up telephone connection, a telephone handset that would nestle in the 'bunny rabbit ears' on the back of the Silent 700, and knowledge of the various telephone numbers, system settings, user IDs, and passwords required. With patience and effort on the user's part, typing the correct commands at the correct times, the Silent 700 would be connected to the remote computer and the tasks accomplished. Most of the commands were, to the initiated, logical and straightforward; but to those unfamiliar with the process, the interaction was somewhat baffling, not to mention a great deal of trouble for a novelty that did not appear to accomplish all that much.

In the span of two decades, the progress has been remarkable in terms of number of users, awareness of what 'online' offers, and ease of access. The number of users has exploded from about 150,000 worldwide in 1981 to an estimated average of more than 80 million. If the trend continues, some type of Internet access will be available in the developed countries for anyone motivated to walk into a public library, a cybercafe, or public space where pay-for-access public access devices are located. In the United States, providers and manufacturers of cable television, personal computers, and handheld devices are in the process of making online connectivity a matter of clicking a 'log on' icon. Given the aggressive moves of companies like Bertelsmann, the situation will be similar in the European Community. For those with the need to be online, the opportunity is now available with relatively little effort.

Diffusion

Awareness of online access has in the last few years diffused into the public consciousness. After nearly 30 years in the shadows, the notion of linking to remote information resources and using a network to communicate via electronic mail, voice, or video enhanced voice is taken for granted. The Giga Group, Gartner Group, Booz Allen/Economist Intelligence Unit, and hundreds of other research organisations have stated a variation on 'The awareness of online access has increased sharply.' Even *USA Today* includes graphics that inform the reader '37 percent of airline travellers want Internet access.' In the United States, most elementary schools are benefiting from the government's effort to connect schools. Distance learning, collaborative communication, and online research are being woven into the fabric of primary, secondary, and college instruction, and large publishers like Pearson plc have reorganised operations to serve this new market. For practical purposes, anyone under the age of 17 in the top five Internet-aware countries (the U.S., Canada, the U.K., Germany, Japan, and France) knows what 'online' means. Their parents, sometimes uncertain about details, are finding themselves pressured into getting with the digital programs.

The community of information publishers has changed even more dramatically. In 1981 only professionals with access to costly systems, knowledge of complex programming languages, and money could create an online service. The first editions of Carlos Cuadra's database directory in the early to mid 70s counted about 500 databases available from about a score or so services. Estimates of the current number of online 'services' are difficult to pin down. If one averages the number of web pages indexed by the commercial services All the Web (http://www.alltheweb.com), Northern Light LLC (http://www.nlsearch.com), and Lycos/Hotbot (http://www.hotbot.com), there are at least 140 million web pages. However, these indexes do not catalogue all levels of a web server, exclude password protected intranet and extranet services, and 'pages' on pay-for-access Internet sites. Hence, known services are but the tip of the content iceberg. The key is that anyone can now be a publisher: services such as Lycos/Tripod, Yahoo!/Geocities, and others allow individuals to be publishers from one minute to the next – even if what is published is a photograph of the family dog.

Changes in 20 Years

What do these shifts during the span of two decades portend? The more important checkpoints include:

- Firstly, the act of connecting is rapidly becoming an unnecessary task. Connectivity is ubiquitous now for certain demographic groups in developed countries. Wireless portable devices are proliferating as their prices drop from thousands of dollars to hundreds of dollars. An 'Internet tone' is complementing (possibly replacing) the standard telephone dial tone as a communications link. Freed from a wire umbilical, new types of cost savings and services have a fertile ground in which to take root and flower. In Tokyo and on the campus of the Massachusetts Institute of Technology one can catch sight of a few people wearing computers. The integration of the access device and the computing device into part of one's clothing underscores how connectivity is part of the environment in certain locales.

- Secondly, the use of information services is becoming increasingly integrated into other tasks. The availability of specialised information services make 'research' a task that requires no special effort. An answer falls readily to mouse click. Idealab grew from one online user's dissatisfaction with the process of looking for a barber in New York City. City Search was the result. Access to government data is a click away on Northern Light. Information about commercial airline flight arrival times requires no voice call to understaffed airlines' customer service lines. Trip.com provides the answers and will send electronic mail to a user wanting precise flight arrival information. With these services, an informed person conducts 'research' as spontaneously as chatting with a colleague at a trade show.

- Third, creating information services has shifted from the defining tasks of the 1980s online business – designing systems, aggregating information, and teaching users what is available and how to use it. Today, the rewards go to the people who can integrate online access into other, often quotidian tasks. For example, Time0 (a unit of Perot Systems) built OrderZone, a back office service that allows a business to simplify its supply chain interactions. The user of OrderZone has an easy-to-use interface to ordering supplies, verifying the order, determining delivery date, and pulling up status reports. In addition, news and product information are available at a mouse click. Companies using the service can reduce their dependence on certain in-house information systems if they wish to move work tasks from full-time staff to part-time staff working from home. What is happening is that online access has worked like a capsule of medication. The system has swallowed the delivery mechanism, and the business end of information systems has entered the bloodstream of human action. An alternative way to think of the change is that online has moved from a marginalised activity to a lens through which interactivity flows. Online, like the automobile before it, has become a natural part of the flow of life in developed countries.

What's Next?

The developments outlined above serve as a jumping point for a bit of *stick-the-neck-out thinking*. Prognosticating about the future of information is nearly impossible. Technology – particularly information related technology – has a way of moving rapidly in often unexpected directions with unknown consequences. Who could have predicted that violence in online games would induce students to recreate it in an American secondary school? Who could have anticipated the emotional response of a salesperson when his personal computer crashed during a client information lookup in an airport? Despite the thin ice upon which the prognosticator must tread, several observations can be offered:

Firstly, software will get 'smarter.' For certain functions, layers of software will work like silent helpers. Word processing software can already correct simple transposition errors, changing 'teh' to 'the' on the fly. With the more powerful personal computers, software can take advantage of the hardware to exponentiate behind-the-scenes activities. For example, dMars (Distributed Multi-Agent Reasoning) from the Australian Artificial Intelligence Institute performs a wide range of observation, matching, and look up functions so that a person looking at a web page can, if he or she wishes, access similar sites, see information that similar questions located, and be alerted to other supplementary material. dMars is one example of how a type of software pioneered by Firefly Network and Net Perceptions can be extended with more proactive, 'automatic and intelligent' services.

Secondly, content will be enriched with non-text 'objects.' So far, information has been embodied in text, particularly in the pre-web world. Today's online users want access to audio, image, and video content. The surge of interest in audio content, particularly in the MP3 format, is a harbinger of what will happen in image and video content domains. The barriers to online access for non-text content will be lowered. Users, regardless of age, respond to audio, image, and visual cues. The pressure to deliver what many users want means that these huge and largely untapped pools of content provide a potential bonanza for companies or individuals who crack the many difficult problems associated with automatic indexing, formatting, and presenting of digital video streams. The diffusion process will follow the well-worn path of early adoption by companies with sufficient resources, followed by a broader diffusion over time.

Thirdly, the database will become the centrepiece of online services. Database technology is not as exciting as some other disciplines in online. Despite their seemingly dowdy exterior, multi-object databases are the engines for interactive web sites. The database technology is moving along two paths which will intersect at some time in the future. One path consists of two well-established database models: the flat-file model and the relational model.

The second path consists of object-oriented databases. Objects are often a collection of the data that constituted a record in a traditional database. However, the advantage of the object model is that a higher level of granularity is possible than that offered by individual fields in a series of tables. The benefits will range from reduced programming cycles to more intelligent databased applications. The merging of the two paths can be discerned in the most recent releases of IBM's DB2, which has a series of stable tools to blend relational and object-oriented functions. No one type will fully disappear, but over the next few years, more emphasis will be placed on manipulating objects in distributed network architectures.

Impacts?

Smart software, objects, and databases mean nothing without applications. In closing, it may be useful to describe how these components might be exploited by a 'total network' or a 'ubiquitous network'; that is, a network which does not require logging on because users are already connected.

Smart 'Monitoring'

Much press has been given to the development of systems where a bar code scanner monitors what one places in a refrigerator, or where a central computer monitors what one has on hand and spits out recipes optimising the available food ingredients. An interesting option, perhaps, for those in the market for a new refrigerator. More useful will be a wide range of wireless devices that protect our well-being by, say, allowing the elderly or infirm to be connected to medical support systems. A change in a particular reading can be a signal for a change in medication; an unconscious person could get immediate help through a wearable device with a messaging module. There will be those who feel that such devices intrude on individual freedom. However, the use of persistent connections and the bits and pieces necessary to perform useful monitoring activities offer a number of public safety and health care benefits which it would be irresponsible of us not to pursue.

Investor Safety Net

Smart software can work wonders today. But even the most clever routines from data mining companies require human intervention. Flash forward three years to a computing device that learns what its operator does. Using software that can make sense of a user's actions, our computers perform routine tasks and works in harmony with our actions. A person looking for a particular type of investment can rely on software that monitors actions, tracks gains and losses, and 'suggests' better investments. (Now, will the user take the software's recommendations?)

Metrics

There is one other interesting aspect to the placement of online in the mainstream of personal and professional life. It will become more important to quantify the payoff from investments in electronic information infrastructure. With the rich choices online engenders, pivotal questions and equally important answers to these questions lie just behind the digital display:

1. Calculating the value-velocity ration. An investment in online technology must reduce the amount of time to accomplish one or more tasks and return a measurable increase in 'value,' as determined by cost or time analyses.

2. Tracking cost savings in order to create a basics-to-technology investment ratio. The goal is to allocate about 30 percent of each budget dollar to new online technology and infrastructure in any fiscal year. Most organisations operate on ratio different from the 70:30 balance.

3. Setting up audit mechanisms to measure performance against the criteria summarized in the table below:

Category	Goal	Comment
Responding to internal customers	Faster Higher satisfaction	Measuring network investment for an internal constituency
To customer	Faster turnaround Increased sales Higher satisfaction	Payoff requires two axes: actual payback plus a 'satisfaction' metric.
To vendor or partner	Faster Lower costs Reduced turnove	Costs should be held flat or decreased. More critical is the retention of key suppliers or partners
Costs	Reduce over time	Overall, information and online technology must reduce costs, other-wise, traditional methods are 'better'.
Payback	Increase revenue Reduce customer turnover	The company or organisation must 'grow' its cash reserve, revenue, surplus, or other 'cash asset'.

Information Metrics

A Look Ahead

In closing, I pose five tangible results from the mainstreaming of online:

- Firstly, near-universal access to a network support 'environment.' Controversy about the intrusion of online into everyday life is a concomitant to pervasive networking.

- Secondly, increasingly easy access to online services. Voice inter-action will lead to subvocal interaction on an accelerated time line. Software is available that allows a personal computer to respond to voice instructions. Rapid improvement in this area is going to transform a person's interaction with a computer.

- Thirdly, finer and finer gradations will exist among computer spe-cialists at the same time they will cluster in a handful of locales. An imbalance in computer expertise will become a major concern in certain nations.

- Fourthly, greater distance between computer 'experts' and compu-ter 'users' as computer access and online services 'disappear' from everyday life. Because the digital environment becomes part of the everyday environment, it will become more and more difficult to 'see' the new world of analog and digital realities, and individual dependence on experts could prove a serious challenge – just as the innards of cars have become so chip-ified that do-it-yourself troubleshooting and maintenance are almost out of the question.

- Fifthly, significant issues of control, governance, and privacy are just over the horizon as previous definitions of permitted and pro-hibited activity prove inadequate. The best and the brightest minds will be needed to sort out the legal and regulatory ramifications of all the brave new things we are suddenly able to do with so little effort.

In closing, a myriad possibilities arise from the mainstreaming of online. The challenge is for the public and private sectors to invest wisely for sensible, beneficial applications that will enhance our common quality of life and solve the problems we have wrestled with for so long.

12

A Cup Half-Full

Ian Brackenbury

We might sometimes, like playwright John Osborne, look back in anger, but we usually look forward with hope. Each new year and new decade arrives accompanied by plans, resolutions and the hope that good things are around the corner – and the start of a new millennium carries a great deal more promise.

The globalisation of society, driven by mass communications and the breaking down of barriers, means that we are moving from a world where independence is valued, to one with an increasing focus on inter-depend-ence. We can envisage a global 'village' where an increasing proportion of the population can participate in what matters to them, where the freedoms we take for granted in the Western world are more widespread, and where choice – of work, and play – is dramatically increased. People will be able to spend more of their time on what's important and relevant to them, and in the next few decades they will reap the benefits of the information technologies that are evolving and improving in computer laboratories around the world today.

e-Business and e-Pleasure, e-Work and e-Play

At one time, when we discussed e-numbers, we were talking about the list of colourings and additives to food and drink, likely to have disastrous effects on hyperactive children. Today, we have a whole e-vocabulary, and e-business (communicating and trading via the World Wide Web) has become big business. The technology underpinning the rapidly emerg-ing e-society can be seen in five emerging themes.

Ubiquity

The massive growth in the use of PCs, laptops, palm-sized digital assist-ants, and mobile phones means that a large fraction of the world's population can communicate with each other electronically. The mar-riage of the Internet with the rich information technology infrastructure laid down over the past few decades enables people to talk to each other – electronically – any time, anywhere. Virtual teams can work together across the globe, without ever meeting. The staggering growth of mobile phones, and the technology that allows users to access the Internet, send e-mails and faxes from their handsets, means that business people are rarely out of contact.

Technologies are converging to meet users' needs to move voice and data, and the emergence of worldwide standards based on Internet Protocol (IP) is accelerating this trend. IP, allowing the convergence of voice and data transmission over the same carrier infrastructure, reduces the cost of bandwidth. As costs come down, higher bandwidth applications, such as full motion video, graphics, and large file transfers are becoming much more affordable, and more widely used.

Third generation networks, based on the technology standard UMTS (Universal Mobile Telecommunications Standard) will allow mobile phones to carry out tasks we couldn't have envisaged just a few decades ago. These new networks will be compatible with those in the US, unlike the current GSM, and roaming agreements worldwide will ensure we can always stay in touch.

Of course, it's not only work that benefits from this technology. Families can send photos and videos across the Internet, people can search out the best car deals, buy goods, order services, join common user discussion groups, or just send electronic letters to their favourite pop star.

Technology is driving remarkable social advances that will enable many more people to work from home, or wherever is most suitable, and to increase personal productivity, pleasure and the overall quality of life.

Value-of-time

The most often heard complaint, or excuse for not doing something today, is that we don't have enough time. Unlike any other period in history, the majority of people in Western societies find they are habitually being beaten by the clock.

Information technologies can – if we are prepared to pay for it – give back some of that time, and help us manage our precious hours more effectively. For instance, your software 'agents' can answer questions for you by searching the Internet while you're doing something else. While your agents check and filter data for you, you could be enjoying real-time collaboration with colleagues around the world – discussing projects together via videophone or a web-enabled TV.

Hours spent on cramped aircraft for short business meetings across continents will be a distant memory. Business people will have video-audio speech contact with their colleagues, enjoying face-to-face electronic meetings. The only element missing will be the human smell, but after an all-night transatlantic flight and a hot taxi ride, that may not be seen as a deprivation.

Instant Payback

Future generations won't accept the old 'allow 28 days for delivery' any more. When consumers can search products on the web, order them online, and pay electronically, they see no reason for a delay in delivery.

We've gradually accelerated the buying process from ordering via the post, to the telephone, to online – decreasing the actual processing costs and reducing the waiting time. People who work in the manufacturing industry have becomes used to the idea of 'just in time' deliveries, and we may be moving to a society where everyone accepts the same principle. The gap between deciding you want a particular item and getting it will be measured in hours rather than days.

Usability

Tied into the notion of Instant Payback is usability – the expectation of customers who want everything to be 'plug and go'. People will not tolerate computers or any electronic devices that are difficult to use.

This doesn't mean that every product has to be designed with a typical five-year-old in mind, but it does mean that devices have to be simple to use, and any instructions or training has to match the perceived difficulty of the task. There are many people, for instance, who are not happy using a keyboard. They would perhaps prefer a PC they can talk to, one that will recognise their voice, and follow spoken instructions. Customer-proven design will come much more to the fore, especially as companies recognise the growing need to market to an audience of one.

Information about cultural predispositions, and personal preferences will shape the products of the future. Devices will also need to be multi-tasking. People won't want to buy separate pieces of equipment to print, scan, copy, type, view and store information. Multi-tasking machines will replace what we now have, becoming smaller, but more powerful, and simpler to use.

There are two ideas here that I'd like to present.

1.   Computers will combine tasks so that they become integrated 'communication centres' for work and home, in public spaces and as part of drop in centres in business and utilities.
2.   Computers will vanish inside appliances, so that instead of having one PC that is PC-sized and does all the personal productivity and web-accessing things that PCs do, we'll have several digital appliances, where the shape, size and colour match what it's for.

You may have a 'social calendar' for the family, and another for work. It will integrate a number of calendar-related functions – including synching up with the rest of the family and the rest of your work team. It's not going to do faxes or heavy duty word processing, and it will be very light, slim and long-life battery operated.

You may have a Home Server that archives documents, calendars, and looks after shared family (or work) equipment like printers and scanners. A powerful, large-screen laptop may be your personal machine for serious computer work, and you will have two or three PDAs scattered around – one in the car, one in the kitchen, maybe one in your purse or wallet.

The integration comes from their connectedness, so they keep in synch with each other. They never lose information and, like phones in most households today, there will be several of them in various shapes and sizes.

Fashion

When compared with the inflexibility of large-volume manufacturing lines, or agricultural investments in crops and herds, the online information services are infinitely malleable. It takes almost no time to reinvent products and processes relating to customers.

At the same time, comparison-shopping has never been so easy. People can shop around for books, flowers, CDs, pensions, holidays, and cars, for instance, in a matter of minutes, without having to walk down a High Street or open a catalogue. The actual differences in quality and price will be subject to free market rules based on pretty much complete and timely information.

So, the differentiators will often be the ones of perceived brand value and fashion – very subjective, very ephemeral, and for those suppliers who hit the sweet spot – very lucrative.

What's Driving This New Society?

There are several trends that mean we can predict the e-society with some assurance:

- Electronic leisure, especially among the young, and particularly where fashion dictates it's cool to be online. This is matched by electronic business, especially to increase the size of potential markets, to enhance relationships with customers and to improve the efficiency of dealing with suppliers.

- It's rare for a business leader today to offer a card that carries no e-mail address. It's equally rare for a teenager to be unaware of the attractions of electronic chat.

- Personal management, which will be very influential when it becomes ubiquitous, is another driver. We expect software agents and the use of multimedia real-time collaboration to become as normal – and a determinant of business and social processes – as the telephone is today.

- The transformation of government into electronic organisations is as much a symptom as a driver of the online society. An e-government is one which opens up electronic access for its citizens, allowing, for example, online motor registration, taxation information, planning permission and electronic payment of parking fines.

The Enablers

At the two ends of the scale are *pervasive computing* – with embedded microcomputers in appliances, such as cars, fridges, and other household appliances, plus heating and lighting systems; and *deep computing*, with megacomputers and their really serious instantaneous MIPs (million instructions per second) for solving really complex problems in a range of industries. For example, better optimisation of raw materials, simulating the effects of new manufacturing plants on the community and the environment, and making better weather predictions.

As we do more and more of our work online, the benefits of Massive Online Hierarchical Storage will become obvious. Storage will be so cheap, the phrases 'back-up' and 'disk-full' will disappear. Users will never have to throw anything away and won't have to select what to save and what to delete. Good records can be maintained of all transactions, queries, searches and conversations – not just for business, but for all of us.

We will expect bandwidth to be fit for purpose – so information is delivered where it's wanted and with the right quality of service. This would mean guaranteed low-latency, high bandwidth for video; low cost for next day delivery of email, with bandwidth priced and stratified depending on what you want to do and how much you are willing to pay for it.

Open standards for software will put the focus more on applications and solutions, and less on celebrating how complex it is to run these beasts and how clever we are in making them dance. Software agents will filter and condense the information you need, spending time carrying out searches so that you don't have to. Real-time collaboration will allow people who work and live on different continents to bind together in a similar way to the tight-knit communities of several hundred years ago, when the commonest form of communication across any kind of distance was to shout.

We are also seeing new models for business. The traditional way of running a company is being turned on its head in the digital world. We're becoming used to buying, selling and auctioning goods over the Internet, and self-help trading will continue to grow. Customers will be able to order the goods they want, pay electronically, and track delivery via the web site.

The electronic supply chain is much stronger than the old-fashioned paper chain, and we are now seeing electronic relationships building up between suppliers and organisations that may never have met.

It's been said that the great thing about electronic trading is that 'no-one knows you're a dog on the Internet'. Equally, nobody knows whether you're a multinational, or a small, family-run firm. The Internet has provided companies with a level playing field, and in the next decade it will no longer be the survival of the fittest, but survival of the fastest.

A Better World Because ...

Computers are becoming cheaper, smaller and faster, networks are becoming cheaper and ubiquitous, software is becoming more standardised and simpler. The future of information is looking very good indeed; in fact the information technology industry is on an unstoppable path to continue improving year after year.

However, it's not until enough people are connected, enough people are at ease with being online, and when the improved power of the computers and networks translates to easier-to-use, more flexible systems, that the real value of the ever-improving technology becomes apparent.

When this happens, a complete transformation of entire industries, changes in their relationships with customers, suppliers, and the community within which they operate, takes place. For example, education and training will be a constant process for life, with tailored programmes for each individual; in health we will have more preventative as well as recuperative maintenance, and we will balance the use of natural and non-renewable resources.

We will expect, and adapt to constant change – in Government, justice and law enforcement, transport, commerce and the community.

A Different World

Social change, enabled by IT advancements, will give people more freedom to work and play as and when they wish and give them more choices over what they buy and what they do.

Information technology will greatly reduce the 'makework' and administration of both government and business processes, so that people have more time for the things that matter to them. Through improvements in the quality and timeliness of education, health management, transportation, law enforcement, the global community will be greatly enhanced.

The future world will be a place where information spreads rapidly, and much of it is free. Diversity of views and feelings will be expected and tolerated. It is indeed a cup half-full.

13

Back to the Atelier: Academic Entrepreneurs and the Future for Information

Frank Colson

It is not too bold to argue that thousands of small ateliers rather than a few major production line operations will lie in the forefront of information engineering, publishing and academic activity in the XXI century. These ateliers may be loosely defined as studios housing scholars/artisans, engineers and writers working on cognate subjects whose boundaries are defined by no particular technology. But all are concerned with publishing their work in both digital and print form. Small but ever-changing in shape, agile yet robust, the atelier will become a primary agent of change in the information society of the future.

These ateliers will live within the interstices of more traditional enterprise and will be heavily dependent upon electronic media scholars, and by so doing they will stand at the forefront of knowledge. The major learned institutions and multilateral corporations that dominated information engineering in the last quarter of the twentieth century may only survive by devolving many of their functions to the ateliers. The challenge for today's universities and public corporations will be intense because the atelier will explore new perspectives, foster new academic disciplines, implement new connections and create new products. It will draw on a wealth of knowledge unbounded by the overlay of hallowed administrative traditions and bureaucratic controls. Under the guise of ancient corporate and academic garb, ateliers might well generate new wealth, often in a manner that is unexpected yet swift and efficient. They are burgeoning daily. Because of their lean and flexible structures, ateliers will take on many of the information engineering activities of major corporations. For businesses this devolution will be natural, as production processes are taken over by the ateliers. Their development and growth will possibly mirror those groups in Silicon Valley which have contributed to the spectacular performance of Internet companies's shares on the Nasdaq. For those top-down driven universities bent on conforming to state-driven agenda, they will be a significant challenge.

Today's ateliers are no newcomers to the scene. On closer examination it would appear they resemble the workshops and enterprises of the scholar-printers of the fifteenth and sixteenth centuries. Like those sixteenth century workshops examined by many scholars, they comprise librarians, writers, editors, publishers, printers, agents, and distributors.[1] Like the bookshops of old, they do not discriminate between professional scholar and the reading public. In that sense their reach is wider than that of the fledglings of Silicon Valley. Just as in California, today's ateliers might well be located at the seat of a thriving university community. Nonetheless, they will not be parasitic on academe but will embody intensive collaboration within the larger world of research, fostering intensive interchange between people of learning and enterprise. The ateliers may well confer their own rewards and privileges, create an economic alternative to employment in academe, be engaged in far-flung enterprises, and have a decisive impact on the world of enterprise outside the university.

A caveat is in order. Like its fifteenth century predecessor, the atelier cannot be subject to the confines of either university governance or state regulation. They would smother it. Just as Papal regulation stifled the growth of printer workshops in sixteenth century Rome and Venice, so state regulation of university education, albeit in the surrogate forms associated with Fordist productivity measures, will cripple fledgling ateliers. For such measures will invariably fossilize around hallowed canons and disciplinary organisations – a patriarchy of privilege. By binding scholars to rigid accounting structures derived from an earlier and Fordist age they will insulate them from the interplay of a variety of different stimuli. As guild-like regulations are invoked by institutions desperate to find rationales to manage the growth of knowledge, the ateliers will flee to seek support from private, corporate and even state patronage. They might even take over simple services for the commercial community. Unfettered by the archaic mercantilism which governs Treasury control, the ateliers are 'mixed economy enterprises' working with a variety of enterprises, they draw on their wealth of experience – and are free.

Perhaps a practical note is in order. The ateliers we have in mind are operating in response to the many markets that have become available to knowledge engineering. The atelier's economic activities therefore involve all forms of publishing and consulting, but stresses the design and preparation of bespoke indexes and classificatory systems. In technical terms the ateliers will be involved with the design and implementation of the new software devices which can mimic the ways in which we understand the brain to work. They will be inherently concerned with the production and documentation of all forms of management systems, especially as these exploit the freedom offered by intranets. This is because each activity within an atelier has its market, so the management of the atelier is itself an intellectually creative task which spawns its own agenda, devices and publications. Since innovation in any one area of its activities will affect all others, the atelier is also a learning and teaching environment with a 'flat' management structure. If it fails to devote at least 25% of its time to teaching and learning which constantly upgrade staff competence, it will rapidly cease to exist. Such teaching and learning can be undertaken because the atelier has the means at its disposal to protect its intellectual property. In contrast to Fordist organisations in the worlds of academe and the public corporation, the atelier will also remain an open organisation.

Today's atelier will also foster and feed on the cultural metamorphosis which is currently underway and which has been immensely stimulated by the digital revolution. The impact of this revolution has a certain intriguing resemblance to the galaxy of developments in the organisation of knowledge which followed Gutenberg's invention of movable type. As a wealth of recent reflective scholarship has pointed out, these were far more complex and vastly more powerful than had been imagined. The explosion of bible printing encouraged the rise of a powerful and literate lay establishment, and consequently undermined ecclesiastical preeminence. The standardisation of equations, diagrams, tables, maps and charts that was integral to the printing process contributed decisively to the diversification of universities away from the traditional faculties of law, medicine, theology and philosophy towards mathematics, engineering, biology, physics and astronomy. Above all, the business of standardisation led inexorably towards the creation of the critical edition with a decisive impact on theological and historical scholarship. The question 'What is evidence?' could now be asked with some rigour. And the investigation of this question lies at the soul of the atelier.

The current revolution extends the developments of the sixteenth century from letterpress print media to multimedia. Though it is still at an early stage, evidence provided by movement, sight, and sound can now be subjected to the discipline provided by print. They will become more rather than less available, because they can be accessed and therefore analysed in a far more systematic way than was previously possible. It is enough to remember that the new information technologies have begun to displace the primacy of words as theatres of memory in favour of images, sound and gesture. Though someone browsing the Internet would scarcely be aware of this, the digital technologies enable the codification of signs and symbols on a scale that was previously unimaginable. It might be remembered that until the advent of the printing press, sound and gesture – oral traditions – represented ideas and transmitted traditions.

The advent of innovatory hypermedia devices which enable the identification of individual hyperlinks follows on the development of ever more rigorous data mining search engines and open hypermedia systems so as to stimulate the identification, collation and classification of authorship and inference in digitised movement and sound to a degree that was previously inconceivable. Given current developments in neurological modeling, future ateliers could be concerned with the vital task of identifying and stamping authority on the retrieval and collation of multimedia information. This concern with authority derived from evidence will be central to the preoccupations of those working in the ateliers.

The need for this authority is essential because while the digital revolution has made scholarly resources far more accessible than was previously the case, the proliferation of variants of any single document that can be published makes it essential to establish authority. Who wrote what and when precedes why! While early ateliers resembled scriptoria to transcribe and transmit the canonical texts of our cultures according to specific rules, the advent of more rapid communications meant that the image of the text, as well as the text itself, could become available for scholarly comment. Ateliers have rapidly emerged to provide more effective text-searching devices to enable works which have been transcribed to be more widely available to the reading public in general. Early ateliers simply transcribed common reference tools such as lists of telephone subscribers and dictionaries: the obviously marketable examples of letterpress technologies. Given the vertiginous expansion of the resources available on the Internet, canonical texts and images have been made available to scholars on an unprecedented scale. Even though the fundamental research in this area has been undertaken, implementing efficient search devices for image, sound and moving image will remain areas of active research for some time to come. Identifying, documenting and exhibiting 'authority' in these areas will be the principal challenge of the early 21st century. Ateliers will feature largely in this area of activity, so further description might pay dividends.

So long as text-processing remains the paramount form of transmission, the problem of authority can and will be resolved. The work of the Text Encoding Initiative [TEI] has built upon the printer's use of the Standard Generalised Markup Language [SGML] to provide for the effective transmission of centuries of editorial experience to the new world of the electronic text.[2] The SGML browsers produced by many groups working with the TEI are now in general use throughout both the academic and commercial worlds. This is not surprising since economic considerations alone dictate that the publication of major editions will be primarily electronic, with paper-based versions playing a secondary role. Individual libraries will simply create their 'own' printed versions for internal use. This is hardly surprising given the rising costs of production and distribution encountered by academic and technical presses. What is interesting is that the development of the TEI itself is likely to remain the business of ateliers, relying on a mixture of private and public funding, scholarly research and publication. Clearly, major editions of canonical works will continue to merit transcription: debate on the myriad changes in a Shakespeare manuscript can deepen and can lead to dramatic shifts in direction as scholars hone various tools to provide ever more sophisticated 'renditions' of a given text. At the same time search engines developed for scholarly purposes can be effectively translated to the commercial world to more prosaic but no less challenging tasks.

While the TEI does seek to tackle canonical works, it has two limitations – the number of works themselves, and the limits of text transcription. Both affect authority. The issue of authority will become more important as the massive expansion of the second hand book market created by Internet technologies enables collection and digitisation worldwide of works that had previously lain abandoned in dusty shelves. Who knows just what editions might lie 'undiscovered'? A caveat is in order here since the revolution will not merely encourage the republication or edition of works of quality; it does not discriminate, so new forms of unfocussed mystification will continue to emerge. The canonical texts will no doubt be enlarged in number as new versions are discovered, and the business of the ateliers will be to constantly indicate the extent of their originality. This process – which inevitably includes significant software development – will involve the entire range of staff within the ateliers.

The ateliers will not merely be concerned with quantity, but with authority. As the digital revolution unfolds, the search for technical and authorial standards will be of paramount importance. Otherwise there is the lurking danger of a Gresham's Law of information, especially where essential scholarly tools are concerned. The more information abounds, the more we must have reliable and uniform tools with which to work; calendars, dictionaries, reference guides, maps, diagrams and charts are of vital importance if time, that most vital of resources, is not to be wasted. The advent of data in multimedia forms makes this ever more pressing; for instance commercial pressures will naturally spur the development of more powerful multimedia search engines, but will not necessarily aid the production of high-quality scholarly tools such as thesaurae. At the same time, the pressure to distinguish the original from variants will demand the production of devices which enable scholars to see exactly when a publication has been altered. There is ultimately no escaping the need to distinguish and classify the original, whether as words, shapes, hard-coding, or alteration, in whatever form or format. Initially in software but subsequently in hardware, digital technologies can be harnessed to encourage a far more systematic because more wide-ranging study of authorship than has previously been the case.

At the same time the dramatic fall in the cost of digital storage media and the extraordinarily rapid diffusion of more efficient communications mean that the image of the text, moving image, high-quality image and sound are now available on an unprecedented scale. This will have an unprecedented impact on the ateliers – the knowledge engineers of the future.

At a most basic level, the work of an individual author or atelier will be better identified, because the text can be seen as it was meant to have been read, together with earlier versions. This has vital implications – editions of images and moving images can be created so that scholars can reconstruct the making of a work, and hence examine the author's intention in a more systematic manner than has previously been possible.[3] How have authors, whether of images, moving images, sound or text 'read' their sources? Previously we have had to work with inference, systematically building up a 'sense' of a writer's intentions – digital versions can enable us to understand a given reading of sources and the context in which a 'public' understanding of an event has emerged. Anyone doubting the economic importance of authority has only to consider the implications for Intellectual Property Rights – these are massive, and in many respects the jury is still out on issues of the 'origination' of important documents.

Examples abound, for the process of mirroring the business of writing has now been demonstrated a number of times, most notably with the publication of Art Speigelman's *The Complete Maus* in which the author allows the reader to take apart Spiegelman's process of constructing his narrative and daily perspective of the Holocaust, in particular his relationship with his father Vladek. The end of Brazilian slavery was itself a human experience as defining as the Holocaust in its impact on a people and their erstwhile masters. Slaves had comprised as much as 10% of the population in 1872, and at least half a million were freed in May 1888. Typically, the explanations which have been given for the sudden collapse of slavery in late nineteenth century Brazil drew deeply on those provided at the time of the passage of the 'Golden Law' of 1888 which brought slavery to an end. Men, writing mostly in newspapers, sought to explain the liberation of over half a million slaves by a single law, passed in the course of less than a month by seeing it as the triumph of white European modernity over an African colonial past. Such explanation privileged the rhetoric of the country's white and propertied elite. This elite emphasised its concern to be part of a world which included the USA, Argentina, on the one hand, and the world empires of Europe on the other. But do these explanations adequately describe the options available to those in power at the time? One must examine the documents as 'read' by those groups. Ten years ago this would have involved a substantial visit to the country, and the publication of a substantial number of commented 'facsimile editions' – it would be an economic and physical impossibility.

Now at least the print media are available in electronic form. Thanks to the remarkable initiative of the Conference of Research Libraries (CRL), the documents perused by senior administrators and the admittedly minute reading public of the era are available online.[4] Via a search of reports from provincial authorities (Brazil was, at that time, a unitary state) it becomes possible to see that as late as 1884-7 slaves were still seen as major assets whose value was high, and slaves were often noted as purchasing their freedom with their own savings. Browsing through the pages of the Presidents of the Provinces for 1884-7, it is abundantly clear that slavery was very much a going concern, and evidently existed through large swathes of the countryside. Municipalities located many hundreds of miles from the capital, Rio de Janeiro, may well have depended heavily on slave labour to cultivate sugar and cotton, often using the most advanced technologies which the nineteenth century could offer. How else can one explain the fact that by 1887, 2,211 slaves and their families in the vast Northern Province of Maranhão contributed some £10,985.10s and 3d (in 1888 exchange) towards their emancipation?[5] In an age when slaves were not normally seen as earning wages, this was no trivial sum. Clearly then, the unconditional liberation of slaves in 1888 was by no means a foregone conclusion, but rather one of many options suddenly taken by the most powerful men and factions in the period. While it may have been expedient to cloak abolition as a progressive act, it is equally clear that immediate and unconditional abolition would have destroyed many fortunes, jeopardised many interests and altered the relations of man and task in the country at large. Viewed as central to commercial life of Maranhão in the 1880s, twenty years later the black 'embodied the main drawback of the progress of this State'. As far as the official world was concerned, the end of slavery was then more a question of achieving progress by eliminating the colonial/black traditions of society, discarding a heritage in the hope that something more modern might result. It was not surprising that an article in the weekly *Veja* magazine captured online noted that the census records and social surveys of the late 1990s still found those of black [slave] origin as among the poorest and most deprived of Brazilian society – the fourth world.[6]

Such a powerful finding warns us that it might be wise to delve just a little further into the issues which have informed this piece of research. Instead of tramping dozens of scattered libraries and repositories, colleagues can now search a collection which is more complete than at any time since the events of 1887-8 which witnessed the end of slavery. Furthermore, the same technology that makes hundreds of government documents available also provides facsimile reference sources such as directories, guides and newspapers.[7] Given the local development of search instruments, the world of official and public documentation emerges to inform the scholar of the year 2000 in much the same way as it informed the politician of 1887-8. The convenient authority provided by the subsequent inference of the elites – the current synthesis – can be set in context. Electronic media handled in this way provide a powerful incentive to revision, often empowering student writers in significant ways and ensuring the popularity of the atelier in the process. As the work on *Maus* also demonstrates, the work of the author is vigorously exposed.[8]

In the atelier, scholars develop tools which allow them to mirror the various ways in which the information which contemporaries had to hand might have been viewed by those who were able to obtain it. Who read what, and in which way? Systematic cross-collation of the various data indicate that indeed slavery's peaceful end was seen as politically feasible but its aftermath remained as obscure and unknowable. For few societies had survived the end of slavery intact, and the experience of Haiti and the US Civil War were seen as catastrophic, the one as social and economic havoc, the other with its estimated one and a half million dead as the war to end all wars. Men walked appalled by possible consequences.

Let us delve a little further. It is perfectly possible that the horrific alternatives of late-1887 Haiti or a War of Secession can be mapped. The newspapers record that slaves were fleeing the plantations of the São Paulo area as they were being cajoled to work in neighboring Rio de Janeiro or Bahia and Maranhão. These events, juxtaposed on alternate columns in the capital city's newspapers, seemed irreconcilable. The challenge to the traditional synthesis seems clear. Twelve months before abolition few dared hope for a peaceful outcome because the scale of slaveholding continued to be large until a few months before the end in May 1888. The officers of the English banks variously estimated the amount invested in slaves as ten times the annual revenue of the government. Slavery was more ubiquitous, more pervasive than had previously been imagined. Its end could not be imagined as a simple administrative act. Nor was it so imagined.

The extent of the imaginative leap needed to understand May 1888 can now be envisaged for the digital revolution enables the historian of late nineteenth century Brazil to view, collate, and process the information that was available to, and very possibly read by, the men in power at the time. These men had access to newspapers and official reports – so have their historians. The daily newspapers mapped a world of connections, the inferences to be drawn are a web of cross-references, and these can be identified and collated using 'intelligent agents' created as a scholar reads the sources. These allow cross-references to be stored and available for instant consultation, so that scholars can now inform their inferences rather than merely adduce them from speeches and actions. Scholars can work to process data – in an increasingly rigorous manner – because by doing so they can more effectively establish their sources and document their interpretations. At the same time the scholar can record his inferences via hypermedia links and even compile 'intelligent links' that can record the contending hypotheses which might emerge. Slowly but surely the scholar can reconstruct the theatre within which the minds of the major political actors worked, and echo the inferences which emerge as reconstruction continues. An extraordinarily close reading of elite perception emerges: with the sudden abolition of slavery becoming the least bad among a host of different alternatives at a time when events were moving beyond any possible control, May 1888's terse liberation was the least undesirable alternative – an act of mere public security. Of course it could be couched in the language of white supremacy and regal generosity. This made the loss acceptable to those elites living in parts of Brazil that were thoroughly equatorial and hardly conducive to European migration. They could hope for a whiter and less African world.

Two things are happening here. By rigorous classification, cataloguing and sifting, using a range of search engines, scholars working with developers and publishers are altering perceptions of the past. At the same time, an older form of argument, the counterfactual, is being put with a degree of erudition which is almost overpowering. We are all pointillistes now. Some would argue that this has to be the case.

In addition, the scholarly enquiry has pushed forward the boundaries of software development. It will also radically alter the nature of publishing, supplementing the printed page with the sources from which inference has been drawn, the search and collation strategies which have been invoked, their results and the various intermediary arguments. *Ex cathedra* inference will never be acceptable. To paraphrase an old adage, readers will view the book as the completed palace, the sources invoked, and inferences drawn as the scaffolding, builder's mouldings and temporary constructions. Now the reader will be able to construct their wing and indicate their use of appropriate building materials, consultant architect and suppliers. The hegemony of the author's words will no longer be self-evident and will subsist on power of argument over inference.[9]

Of course the scholar in question, attempting to wrestle with the problem via the serried layers of earlier commentary is thus as hideously exposed as were scholars who sought to integrate all views of the planetary system before the availability of print provided for systematic observation. The new tools to be developed by the ateliers provide us with a way of re-imagining the simultaneous theatre of 1888. And that simultaneity is no different in 1888 to 1999-2000. You can read newspapers across the globe only minutes after they have hit the streets, and witness TV broadcasts from CNN, BBC World and Sky as simultaneous theatre – in which several different, apparently unrelated, actions take place on different parts of the stage at the same time.[10] *Simultantheater* has moral and intellectual implications with which we may be familiar. But they remain strangely fused unless we have the tools with which to classify, codify and create the thesaurae with which to undertake their analysis. And what, if anything, lies on the cutting room floor?

Simultantheater demonstrates the need to identify authority in moving image and sound. The need to identify is not merely technical but has a moral imperative. While the technical developments of the ateliers may well lead to a whole new era in publishing within the world of interactive digital media, they might also address issues of authority which have lain unspoken because impossible to annotate. They have remained on the cutting room floor. The crucial factor is that these developments will not take place in the MGM, nor IBM – relics of the 20th century that they are – but rather in the miniature world of the ateliers, and the cruelly adjacent world of the scholar.

Like the printing presses of the XV century, the ateliers of the early XXI will hopefully provide a crucial component to the management of information in the XXI century. The business of managing multimedia information is in its infancy and will require the development effort, in hardware and software alike, of the ateliers' craftsmen-scholars. They will undertake this because they live to identify and challenge authority – in the shape of opinion. The paradox is that they are subversive of post-Fordist society but crucial to the achievement of the economic and social change which is so much the object of government policy. Yet what will be the impact of their growth on government – not so much the government of business, but on the pretensions of the policy makers of official academe. For both official academe and the public corporation, the creative format of the ateliers of writers, software developers, engineers – scholars and craftsmen – may, like its sixteenth century forebear, be both too convenient and too menacing for institutions straightjacketed by financial formulae derived from another age. They will find the atelier an impossible menace, and seek, by disabling publication in electronic form as scholarly activity, to curtail the scientific advance which it engenders. But through its dynamism, plain tenacity, and ferocious participation at the moral and technical edge of knowledge engineering, the association of authority will be evident for all to see.

References

1   Elizabeth L.Eisenstein, *The printing press as an agent of change. Communications and cultural transformation in early-modern Europe.* (Cambridge University Press, 1979.)

2   TEI developments during 1999 and the fate of future atelier are documented in the Report by John Bradley on the TA Software Meeting at Bergen, Norway http://www.cse.fau.edu/~tom/lis

3   '... putting information in digital form makes it easier and faster to get down to the genuinely difficult work of analysis and synthesis.' 'So, What's Next for Clio? CD-ROM and Historians,' Roy Rosenzweig in *The Journal of American History* 81, 4 (March 1995): 1621-1640

4   Centre for Research Libraries/Latin American Microfilm Project's Brazilian Government Document Digitization Project. http://wwwcrl.uchicago.edu/info/brazil. The work of CRL is in many respects that of an atelier, with indexing, guides, and other searching aids devised by staff on the Project, which is funded by the Andrew Mellon Trust. Typically the pages are produced by a commercial organisation, NetOn-Line Services and Hollyer & Schwartz.

5   CRL Provincial Presidential Reports (1830-1930) Falla que o exm. snr. dr. José Bento de Araujo dirigiu á Assembléa Legislativa Provincial do Maranhão em 11 de fevereiro de 1888, por occasião da installação da 1.a sessão da 27.a legislatura. Maranhão, Typ. do Paiz, [n.d.] Falla que o exm. snr. dr. José Bento de Araujo dirigiu á Assembléa Legislativa Provincial do Maranhão em 11 de fevereiro de 1888, por occasião da installação da 1.a sessão da 27.a legislatura. Maranhão, Typ. do Paiz, [n.d.] http://crlmail2.uchicago.edu/bsd/bsd/392/000042.html

6   Sociedade - Andando para cima, *Veja*, 16/06/1999. '... although they form half of the heads of family in the population they are rarely at the apex of society and more frequently at its base'.

7   The CRL Project houses the Presidential Messages, Ministerial Reports, and the *Almanak Administrativo, Mercantil e Industrial do Rio de Janeieor* (1884-89) which was the administrator's reference guide of the period. These are indexed and searched by facsimile page. See http://wwwcrl.uchicago.edu/info/brazil/

8   The technical difficulties alleged by examination boards in dealing with electronic works presented by students utilizing new forms of hypermedia echo Eisenstein's comment that 'conditions that guarantee speculative freedom are probably related to the 'development of originality'.

9   This use of the CD-ROM to show us history as a complex and active process of construction rather than a simple matter of recovering the 'facts' is particularly appropriate since Spiegelman's original comic book is, in part, a reflection on the profound problems of remembering, reconstructing, and re-presenting the past. In Rozenswieg's review of *The Complete Maus* (CD-ROM) by Art Spiegelman, prod. Elizabeth Scarborough. 1994. (Voyager) 1 CD-ROM, ISBN 1-559-40453-1.

10   Timothy Garton Ash, *Prospect*, June 1999.

14

Why We Need a Science of Information

Keith Devlin

The Coming of Age of the Information Age

A modern company has many assets that have to be properly managed. The most obvious are physical plant, personnel, and the company's financial assets. Management of each of these assets requires different kinds of expertise, which may be provided by the company's regular employees, or it may be outsourced.

In today's commercial environment, and increasingly in that of tomorrow, information is another asset that requires proper management. Of course, information has always been important to any organisation. But it is only within the last fifty years or so that it has become a clearly identifiable asset that requires proper management. The reason for the growing importance of information within the organisation is the growth of computer and communications technologies, and the increasing size and complexity of organisations that has in large part been facilitated by those technologies. It is both a cliché and a fact that information is the glue that holds together most of today's organisations.

Actually, that last metaphor is often apt in a negative way. In many cases, information acts as the glue that causes things to stick fast when it should be the oil that keeps the wheels turning. A familiar scenario in today's business world is for a company to introduce a new computer system to improve its information management, only to discover that, far from making things better and more efficient, the new system causes an array of problems that had never arisen with the old way of doing things. The shining new system provides vastly more information than was previously available, but it is somehow of the wrong kind, or presented in the wrong form, at the wrong time, or delivered to the wrong person, or there is simply too much of it for anyone to be able to use. What used to be a simple request for information to one person over the phone becomes a tortuous battle with an uncooperative computer system that can take hours or even days, eventually drawing in a whole team of people.

Why does this happen? The answer is that, for all that the newspapers tell us we are living in the Information Age, what we have is an information *technology*, or rather a collection of information technologies. We do not yet have the understanding or the skill to properly design or manage the information flow that our technologies make possible. In fact, it is often worse. In many cases, companies are not even aware that they need such skill. Faced with the persuasive marketing of ever-more powerful and glitzy computer systems, there is a great temptation to go for the 'technological fix.' If the present information system is causing problems, get a bigger, better, faster system. This approach is like saying that the key to Los Angeles' traffic problem is to build even more, and still bigger, roads.

In many organisations, the computerized information system, far from being the panacea it was promised to be, acts as an 'information bottle-neck,' that slows up or prevents the information flow it was supposed to facilitate. The solution? Just as the company has experts to manage its other assets, so too it needs experts to manage its information assets. Alongside the lawyers who handle and advise on the company's contracts and the accountants who handle and advise on the company's financial assets, should be the 'information scientists' who handle and advise on the company's information assets.

But there is one problem. There are, at present, no such 'information scientists.' The world of information flow does not yet have the equivalent of a lawyer or an accountant. There is not even an established body of knowledge that can be used to train such people. To become a lawyer, you go to law school and follow a well-established educational path. To become an accountant, you learn about the various principles and theories of accounting and finance. But there is no established 'information science.' (There is an academic discipline called 'information science,' but this is not what is meant here. What is often now called 'information science' used to be called 'library science.' The latter name is more accurate, but as libraries themselves became more electronically oriented, the former name was adopted as being more in keeping with the times.)

While the world of information flow does not yet have a corresponding body of expert managers and consultants in the way that the world of financial flow has financial managers and accountants, the need for such expertise becomes more pressing every day. In response to that need, numerous attempts are being made, all over the world, to develop the underlying 'science' – the requisite body of knowledge – and the appropriate sets of skills. Central to those attempts is the need to understand just what information is, how it is stored, and how it flows through the company.

It's time for the Information Age to come of age.

A science of information will – of course – have to begin with what we know about information. Imagine for a moment trying to take that first step. You will soon realize that the world of information is mired in confusions, not the least being the fact that different people use the word 'information' to mean very different things.

The New Iron Age?

In terms of our scientific knowledge, today's Information Age is reminiscent of an earlier era: the Iron Age. Imagine yourself suddenly transported back in time to the Iron Age. You meet a local ironsmith and you ask him 'What is iron?' How will he answer? Most likely, he will show you various implements he has made and tell you that each of those was iron. But this isn't the answer you want. What you want to know, you say, is just what it is that makes iron *iron*, and not some other substance.

For all that he may be a first class ironsmith, your Iron Age man cannot provide you with the kind of answer you are looking for. The reason is that he has no frame of reference within which he can even understand your question, let alone give an answer. To provide the kind of answer that would satisfy you, he would need to know about the atomic structure of matter – for surely the only way to give a precise definition of iron is to specify its atomic structure.

Today, in the Information Age, we are struggling to understand information. We are in the same position as Iron Age Man trying to understand iron. There is this stuff called information, and we have become extremely skilled at acquiring and processing it. But we are unable to say exactly what it is because we don't have an underlying scientific theory upon which to base an acceptable definition.

To emphasize my point, consider the familiar question: 'If a tree falls in the forest and there are no creatures around to hear it, does it still make a sound?' A common answer is: 'No, the falling tree generates a vibrational wave that propagates through the surrounding air, but that wave only constitutes *sound* in the presence of a hearing creature.' Many people (myself included) would want to amend this answer, saying that the vibrations in the air caused by the falling tree (only) constitute sound if they are of a kind that would constitute sound for a hearing creature, *if one were present*. In other words, it's not the actual presence or absence of a hearing creature that counts, rather it's the *type* of vibration – vibrating air is classified as sound if it is of a kind that would constitute sound for a hearing creature. If creatures with hearing had never evolved on the Earth, then there would be no such thing as a sound, and falling trees would not generate sound, they would merely disturb the surrounding air.

The point is that sound is not the same as vibrating air – or vibrating anything else for that matter. Vibrating air is an absolute phenomenon, that is, it is independent of any observers; sound, on the other hand, is an observer-relative concept. The same is true for information. Information is not an absolute concept; it only arises when there are creatures that can acquire, process, and use it.

The implicit assumption that information is absolute lies behind many of the communication breakdowns and inefficiencies in information flow that we experience in business and in our everyday lives. We will continue to have major – and often costly – problems handling information until we have a better understanding of just what information is.

How do I reconcile my observation that information is agent-relative with the undeniable fact that we have depositories of information – books, newspapers, railway timetables, databases, and the like? The answer is that I am trying to be precise in my use of the word information. Strictly speaking, what you find in books, newspapers, and all the rest are *representations* of information.

Many sources do not bother to make any distinction between information and a representation of that information. In fact, much of the time people use the word 'information' to mean any one of: representation of information, data, knowledge – or information.

The Fundamental Three

The most common way of distinguishing between data, information, and knowledge – the three fundamental notions of the Information Age – is to take *data* as basic and define *information* to be data that has been given meaning and *knowledge* as information that is available (in the mind of a knower) for immediate, purposeful use.

For example, the well-known management consultant and author Peter Drucker has described information as 'data endowed with relevance and purpose.' And academics and management consultants Thomas Davenport and Laurence Prusak say in their 1998 book *Working Knowledge* that 'Data becomes information when its creator adds meaning.' (p.4)

In terms of an equation:

Information = Data + Meaning

When a person internalises information to the degree that he or she can make use of it, we call it knowledge. For example, if I know how to buy and sell stocks and am familiar with some of the companies whose stock values are listed in the newspaper, the information I obtain by reading the figures (the data) can provide me with knowledge on which to trade stocks.

Davenport and Prusak define knowledge in this way: 'Knowledge is a fluid mix of framed experiences, values, contextual information, and expert insight that provides a framework for evaluating and incorporating new experiences and information. It originates and is applied in the minds of knowers. In organisations, it often becomes embedded not only in documents or repositories but also in organisational routines, processes, practices, and norms. (*Working Knowledge*, p.5.)

As an equation:

Knowledge = Internalised information + Ability to utilise the information

Information can be regarded as a 'substance' that can be acquired, stored, possessed either by an individual or jointly by a group, and transmitted from person to person or from group to group. As a substance, information has a certain stability and is perhaps best thought of as existing at the level of society. As such, it is amenable – at least in principal – to being analysed mathematically. Knowledge, on the other hand, exists in the individual minds of people. In consequence, a study of knowledge will involve the techniques of psychology and sociology.

The standard description of the tripartite classification just described is fine as far as it goes, but is not sufficient to support a precise – dare I say scientific – analysis of data, information, or knowledge. For one thing, it leaves unanswered the question 'What is data?' The answer often given, that data is what you find in reports and in the tables, charts, graphs and articles in newspapers is suggestive, but on closer analysis reduces to: 'Data is that which, when endowed with meaning, becomes information.' We find ourselves coming round in a circle. In fact, taken together, the two equations given above serve as a basis for precise definitions of any two of data, information, and knowledge, given the remaining one.

One consequence of that last observation is that even data is not absolute, i.e. it is relative to an agent. This point needs to be stressed, since even among those who acknowledge that information is not absolute (i.e. that what constitutes information is relative to an information processing agent), it is often suggested, or tacitly assumed – erroneously – that data is absolute (i.e. it is not relative to an agent).

In setting out to develop a rigorous, scientific analysis of information – and derivatively obtaining a better understanding of data and knowledge – the first step is to develop formal machinery that can capture the way that words spoken or written, signs drawn, symbols painted, bits stored on a magnetic or optical disk, or whatever can represent information. That machinery will have to be mathematical, since information occupies the same conceptual realm as do the object of mathematics (the numbers, geometric figures, topological spaces, etc.) – namely, the shared Platonic realm where minds meet.

What do I mean by that last observation? Representations of information, by their very nature as *representations,* are to be found in the physical realm. At the other end of the spectrum, knowledge is to be found in the (private) minds of individuals. Information (and the objects of mathematics) can be found half way between those two extremes: It is not physical; on the other hand it occupies a shared conceptual realm. (If it didn't, we could not transmit it from one individual to another.)

(At this point, a side remark is in order. Some commentators have argued that there is really no such thing as information. They often base this claim, at least in part, on the observation I just made that information can only exist in a shared conceptual realm – which they then go on to argue is a nonsensical idea in itself. I have to admit I have some sympathy with such a claim. In a sense, information does not exist. Talk about information is just a way of talking about certain kinds of human – and perhaps non-human and maybe even non-animate – activity. However, it is not only an appealing way of talking about such activities, I think most of us find it unavoidable. Moreover, these days it is a fairly ubiquitous way of talking about those activities, and that alone makes it worthy of analysis. To anyone who insists that there really is no such thing as information, everything I say can be viewed as the result of adopting what I would call *the informational stance* toward certain activities by certain agents, including humans. That is to say, regard all the talk about information in an 'as if' sense.)

In any event, the time has come for the development of a sound mathematical underpinning for information technology, that will provide the information engineer with the same foundation that physics provides the civil or electrical engineer. Let me finish by telling you about one promising line of approach.

The Mathematics of Information

At the start of the 1980s, a small group of researchers in California's Silicon Valley got together to try to work out a new 'mathematics of information.' With the aid of a $23 million gift from the System Development Foundation (a spin-off from the RAND Corporation), in 1983 the group founded an interdisciplinary think-tank at Stanford University: the Center for the Study of Language and Information (CSLI). (I myself started to collaborate with the CSLI researchers in 1985, and have been formally associated with them since 1987.)

The main approach to information pioneered at CSLI was a new mathematical theory called *situation theory.* The initial, groundbreaking work on situation theory was done by Jon Barwise and John Perry, respectively the first and second directors of CSLI.

Situation theory is a mathematical theory – and a collection of mathematical tools – designed to provide a framework for the study of information. The theory takes its name from the mathematical device (a 'situation') introduced in order to take account of context.

The first major publication on the new theory was Barwise and Perry's book *Situations and Attitudes*, published in 1983, the year CSLI opened. In 1991, following several years of intense development of the theory, I published an updated account of the new theory in my book *Logic and Information*. That second book remains to this day perhaps the most comprehensive first introduction to the technical aspects of situation theory.

One consequence of the publication of *Logic and Information* was that a number of individuals in the business world became interested in situation theory and its potential as a mathematical underpinning for a useful theory of information. Some of those readers suggested that it would be useful if I could write a more accessible version of the book, aimed at the business reader – in particular, giving lots of examples taken from the business world.

Eventually, that suggestion merged with an observation I had myself made: with the rapid growth of the World Wide Web, the electronic tools of information processing had been placed into everyone's hands, not just those of the 'information specialist' (whatever that term might denote). As a result, modern society had become far more akin to the Agricultural Age – when everyone needed to know how to tend stock, grow crops, and use agricultural implements – than to the Industrial Age it was more commonly compared to, where the key technologies were in the hands of just a few individuals.

Thus, when I finally did write that second book on the subject, I aimed it at both the business person and the 'average person.' What most people needed today, I felt, was not a technical familiarity with the mathematical theory of information, but a common sense understanding of information, based on that theory – what I dubbed *infosense*.

InfoSense

Infosense is analogous to the 'naive physics' (common sense knowledge of how liquids behave, what happens to an object thrown into the air, etc.) that we use in our everyday lives. Just as our naive physics can be formally grounded in mathematics (more precisely, in mathematical models), so too we can build our infosense from a simple mathematical model of information.

The book *InfoSense* (how could I resist taking that as my title) sets out to develop what I believe is the naive understanding of information necessary to function in the Information Age. It describes the situation-theory-based mathematical model for information and gives examples (taken from the real world) to show how it can be used to make sense of the information exchanges that take place among us every day.

I have no way of knowing whether the particular theoretical approach I adopt in *InfoSense* will turn out to have lasting value – there are several other attempts to develop a theory of information besides the one at CSLI. But I do believe that the key to tomorrow's world will be the acquisition of a common sense understanding of information – an infosense – based on a firm theoretical (and probably mathematical) framework. Today, we handle information with the same dangerous ignorance – a kinder term would be innocence – with which Marie Curie handled radioactive substances. Given that information is just as powerful – and as dangerous – a fuel as uranium and the other highly radioactive elements, we cannot afford to leave it too long before we replace our information innocence with (information) infosense.

Bibliography

Barwise, J and Perry, J. *Situations and Attitudes*. MIT Press, Cambridge, Mass., 1983.

Davenport, T and Prusak, L. *Working Knowledge*. Harvard Business School Press, Boston, Mass., 1998.

Devlin, K. *Logic and Information*. Cambridge University Press, 1991.

Devlin, K. *InfoSense: Turning Information Into Knowledge to Survive and Prosper in the Modern World*. W. H. Freeman, New York, 1999.

15

Net Effect on the 21st Century

Lyric Hughes

China Online

The first half of the twentieth century was characterised by a revolution in transportation; the second by huge advances in communications. We are now at the threshold of a golden age of information, a global Alexandria, which will create efficiencies we can scarcely imagine today.

At the dawn of the 21st century, one could focus on the glaring inefficiencies apparent in an age of transition. Information overload is one example of society trailing technological innovation. Viewed logically, information society overload is a philosophical absurdity: there can never be too much information. The real problem is a lack of information that is organised and relevant to a particular user. Insofar as information over the Internet is 'free' this problem will persist, because it takes time, effort and yes, money, to both organise information flow and create the efficiencies economic imperatives bring to bear upon the limitless storage capabilities of the Internet. Pricing creates rank and order, and serves to regulate and control demand.

As technology itself becomes cheaper and cheaper, there will be an opposing trend towards chargeable information; technology will enable the microtransactions to occur that will form the base of a true market for information. The result will be a democratisation of data that was previously only available to powerful institutions and individuals. Wealth creation will become geographically dispersed and the disintermediation effect so often discussed in connection with e-commerce will have an even more profound impact upon information flow and the use of information by those newly empowered by knowledge.

Another common belief is that the future dissemination of electronic information will be dependent upon current models of telephone and PC penetration. Not only will information be platform-independent in the next century, information networks will evolve differently in different countries and at very different speeds. Hardware barriers will be broken down at a highly accelerated rate throughout the developing world. For example, China's low teledensity and PC penetration rates are generally thought to severely limit the growth potential of the current 2-4 million Chinese on the Internet. However, cable television in China boasts 80 million viewers (1999 figures), which is already greater than the current 66 million viewers in the U.S. Internet access through cable television is clearly the way China will leapfrog into high levels of Internet usage by the high end of its demographic pool.

Huge market opportunities will result in wealthy subsets of what we now call 'emerging economies'. Furthermore, supplying content of a character and quality to satisfy this new group of 'netizens' in their own languages will be a challenge to currently powerful media companies. Ten or fifteen years from today, at least two of the world's top ten media companies will be focused on Chinese language content.

As a result of direct access to information by the average citizen, economists will become more powerful than politicians. As long as the economy is good, who the politicians are simply doesn't matter to most people. In fact, this is the real lesson of the Clinton administration. The current popularity of polling, which has been critical in Clinton's decision-making process, is just another step along a path to a future in which social questions will be resolved through digital referendums. Our politicians may one day be seen as another casualty of disintermediation.

Every economic decision will be global in the 21st century. In a world of global capital markets, investment will flow to the regions that are transparent to the international financial community. The very term 'emerging market' will soon be viewed as a quaint twentieth century term to be replaced by a new worldwide paradigm of transparent and non-transparent economies. Looking into the future, developed markets such as Japan, which fall to create a free flow of information, and whose economic structures remain opaque to the outside world, will find themselves competing against formerly poor nations such as China, whose public policy environment favours transparency. China, on the other hand, has begun to understand that transparency is not only a necessary condition to attract foreign investment, but is also a benign way to root out corruption by disintermediating crooked officials who can no longer claim ownership of special information.

Perhaps ironically, at the same time that intellectual property rights are being introduced (with some vigilance on the part of the U.S.) into the developing world, they will face increased challenges in the developed world. How will information be capitalised if no one owns it and can profit from it? This will be a fascinating question for the great legal minds of the 21st century.

If formerly restrictive regimes can be persuaded to see the free flow of information not as a threat to the power of sovereign states, but as the creator of a viable economic infrastructure and a population incentivised to take advantage of the new economy, the next century will be promising indeed. However, local governments will face new challenges due to re-tail competition through e-commerce. Now that consumers can pick and choose merchandise from a global array of goods, the sales or consump-tion taxes they pay must be competitive not only across state lines but in a highly competitive global market. Resulting decreases in local govern-ment tax receipts will force fiscal reforms on a large scale.

Among the unforeseen consequences of the introduction of the Internet will be the ascendancy of economics over politics, a commoditisation of broadband infrastructure, the creation of a market for digital information, the transformation of developing economies through access to informa-tion and cheaper technologies and pressure on governments to create forward-looking policies in the face of an eroding and newly competitive local tax base. The net effect on the 21st century will be positive for those who can foresee the consequences and make adjustments quickly. Those who cling to the vision of the twentieth century will be the new poor – the Information Poor.

16

Yesterday's Tomorrows

Peter Bishop

'The line connecting the past and the future stretches without end in both directions.' Nilva Relffot, *All My Futures.*
Adamantine Press, 2012

(Young boy's face appears on the screen.)

C: Hey, Pops. What are you up to?

P: Reading the mail. Same old stuff. What's up, Craig?

C: I've got a history project for Inquiry next week. I've got an idea, but
       I wondered if you would help me?

P: Sure, but I can't do it for you.

C: No, don't worry! I think I've got a good idea, but I need some more
       material. I thought since it's about history, you might know some
       stuff.

P: Yea, I guess you could say I know some history, having lived most
       of it!

C: You're not that old, Pops. You didn't live during the Civil War, did
       you?

P: No, you're right. The Civil War was a little before my time. But I
       sure saw a lot of history. Living past a hundred gives you the chance
       to see some changes.... What's your idea?

C: I want to do something about the history of technology. I think I've
       found some patterns that I want to check out with you ...You know
       the long wave theories of change, don't you? How progress comes
       in waves every 50 years or so. Each wave is led by a technology that
       creates productivity in one sector and changes all the others.

P: Do I remember? I used to give talks on that subject back in the 90s,
       before most people even heard of Schumpeter and Mensch. I was
       trying to give people a sense of what information technology was
       doing in their lives while it was still changing. My talk was called
       'The Waves of Creative Destruction'. Borrowed the phrase from
       Schumpeter himself.

C: Yeah, well, now that the info tech wave has pretty well played out, I've been trying to compare it to some of the previous waves, you know, the railroad, electricity, petroleum. I thought it would be neat if I could show some parallels between two of the waves.

P: Well, there's a lot of that. But they're different, too, you know. Nothing changed the world like electronics, though the other technologies had their own day, too. Each one seems to create the biggest change, and then the next comes along!

C: Which one do you think I should choose? I know the railroad really changed the country in the 19th century – opening up the frontier and all. It moved people and things so much cheaper and faster than those wagons did, it really opened everything up.

P: Yeah, the railroad did a lot. Not only did the railroads own and run half the country then, they helped start whole new businesses – like the mail order business out of Chicago and the telegraph which started the telegram business. It was even the reason for the wild west – those cattle drives in the movies were going to the railhead.

C: Yeah, I know, but I was thinking of choosing a little more recent technology, like the automobile. You remember the development of the automobile, Pops, don't you?

P: What are they teaching you kids in school today? When was I born if I'm a hundred and three today?

C: OK, you don't need to rub it in. But you did see some of the automobile society, didn't you?

P: Yes, I did. I was born right at the end of World War II so I saw the second and third phase of the automobile society – the transformation and the saturation. The automobile was really something in 50s. We did everything in the automobile. They had places called drive-ins – two types of places, really. One of them you went to eat; the other to watch a movie. We'd pile in the car on Friday night and your great-great grandfather would drive us to the drive-in movie.

C: Yea, we saw some pictures of that last week. You all parked in rows and faced in the same direction. They also said you had sex ….

P: Hold it! I'm not talking about that with you. How do you know about that?… Never mind, let's get back to your project.

C: OK, so here's my idea. It seems that lead technologies go through three stages – development, transformation and maturation. You lived through the transformation and maturation of the automobile society and all three stages of the electronic society. Except for the development of the automobile, you've seen it all, Pops.

P: Yeah, I guess you're right. Let's talk about the development phase. It was crazy then. Electronics, we called them computers then, really went through four developments – the mainframe, the PC, the Internet and the walkabout.

C: We went to a museum last year and saw some of those early computers. They were in boxes. Why was that?

P: The electronics was bigger in those days. They took more power so you had to connect it to the grid and to the net with wires. Wires were everywhere. They also took up a lot of space on desks and tables – nothing like the wall panels we are using here.

C: But wasn't the net always there?

P: The voice net was there when I was born. We used a separate machine to talk to other people – no pictures, just your voice. The broadcast net was also just coming in the 50s. I actually remember our first television. We had one machine for voice, one to receive broadcast (it was only one way); later one for text called a fax machine and finally the computer. Of course, all that is in the net today, but then it was all separate – different machines for different purposes.

C: Must have been kind of complicated. How did you know when to use which one?

P: Well it was pretty obvious really. If you wanted to talk to someone, you used the telephone; if you wanted to send someone a hardcopy …

C: What's 'hardcopy'?

P: That's when we would print something onto a piece of paper.

C: Why would you do that?

P: Well, before the computer everything was on paper. Paper had been around a long time, and people were used to it. It's actually pretty handy, when you think about it. It's light …

C: Not when there's a lot of it, it isn't! We had to lug books home one week for an experience project. What a pain! Imagine having to carry a whole book back and forth all year long just to read a few pages.

P: Well, that was the way we did it. I used to carry a whole armload of books back and forth to school everyday when I was your age.

C: Why carry anything? Why didn't you just tap in and get what you wanted?

P: You're forgetting that we didn't have the Net then. Everything we needed was printed on something. And you're right. Those books were heavy … But where were we? Oh yeah, so you simply used the machine that did that thing – voice, reception, fax, email.

C: What's email?

P: That was a text version of the messages you leave today when someone is busy and closes their access. Without walkabouts, we could only get on the net when we were sitting in front of a computer, so we would write messages to each other and leave them for the next time the other person got on the net.

C: I can't imagine not having my access open all the time. How would people get in touch with you.

(Woman's face appears on display.)

P: Oh, there's your grandmother. Hi, D'Ann. What's up?

D: Hi, Daddy. Hi, Craig. Sorry to break in like this, but I wanted to ask Craig if he's seen the latest Roto-Tour that came out last Monday.

C: No, Gram, I haven't.

D: OK, here's the link. You can see it when you finished talking with your Pops.

C: Thanks, Gram. We won't be too much longer. Pops is telling me some funny things about computers.

D: Yea, your Pops is the one to know. Some days he would spend all day with the computer when I was little.

P: How's the weather, D'Ann?

D: Still hot, but fortunately, I haven't been out since last Tuesday. We took a trip down to the beach, but it was still yukky.

P: Sorry to hear that. We'll see you soon. Bye, D'Ann.

D: Bye, Daddy. Bye, Craig.

(Woman's face disappears.)

P: OK, back to the lesson. Let me show you what those old computer screens looked like. *(Simulated computer monitor and keyboard appears.)* We would enter information using the keyboard.

C: Did you know how to type, Pops?

P: I sure did. I was very fast. Had to be. My handwriting was so terrible that I learned to type when I was younger than you are. It helped because that was the only way of getting information onto the net.

C: Charlie says he knows how to type. He says he wants to be a writer. He thinks he's so cool, but I don't believe him. Nobody could learn in just a few days.

P: You're Aunt Kate taught herself to type when she was very little. Invented her own system. But you're right, no one does that much anymore.

C: So what was it like in those early computer days?

P: Well, the stuff was really kind of magic in one way, but really frustrating in another. Being able to type words and then go back and correct them right away was great. When I was in school, we only had typewriters and you couldn't really correct them.

C: Boy, you really are old, Pops!

P: Being able to enter numbers and having them calculate themselves; being able to send messages instantly around the world; and then with the walkabouts, doing that from anywhere, anytime you wanted. Each step seemed like magic.... But the stuff didn't always work. The computer would crash a lot. I can see by the look on your face you don't know what 'crash' means. It was a term we borrowed from transportation. Do you remember about 10 years ago when that air vehicle hit the mountain?

C: I don't remember, but Mom told me about it.

P: Well, that's a crash. A computer crash is not as serious – it doesn't kill people, but the computer just stops and you have to start it up again.

C: You mean, the computers were not on all the time? When did you turn them off?

P: We turned them off when we were not using them.

C: But when was that? I mean what if you got a call or needed something. Would you turn it on again? That seems like a lot of trouble. How long did 'turning the computer on' take?

P: Well, I know it doesn't make much sense today. But the computer was something extra in our life then. It's something we used sometimes, like when we were working, but we didn't carry it with us like you do, at least in the early days we didn't. After a while, it started becoming more important as it replaced things like the telephone, the television, the fax – all the communication devices. I agree, it must seem strange to you not to have your walkabout with you all the time, but that's the way life was then.

C: Boy, you're right. I don't think I would have liked it back then. Was it hard?

P: No, I don't think any harder than today, just different. For instance, we would have to turn things on and off when we wanted to use them. Today, you just pop the pizza in the microwave and the machine does the rest. In our day, we had to punch in the time we wanted it to cook and push the start button. We did that for all the machines then.

C: I saw on netnews the other day a house with a two-door refrigerator built into the wall, one door opened into the kitchen and one door opened outside so the grocer could leave the food there.

P: Yes, they're still trying to make that work. They were delayed until they perfected the self-closing door to prevent the food from spoiling when the grocer forgot to close it. It's amazing how smart machines are today. Our computers were really dumb back then, but the walkabout seems to know when you want to take a call and when not.

C: Yeah, mine is still new so we're still getting used to each other, but they say after four or five years that most of the training is over. I can't wait not to have to tell it every thing!

P: Getting your own permanent walkabout is the same for you as getting our driver's license was for us – you're finally recognized as an adult. I guess there are more parallels between the automobile and net society than I thought…. So, have you got enough for your project.

C: Yeah, I think so. I had pulled up some images to use, but I really didn't understand them until you explained what it was like. It sure seemed hard. Did you like living back then?

P: Sure, it's really no different than today. We thought we had invented all this great stuff, which we had, but we had no idea what it would be like once the stuff became mature. All the wires, the crashes, looking for files…. No, I'm not going to explain what a file is!… It's a lot like today. Everybody thinks the latest bio-enhancement is so high tech.

C: But it is! We're changing the world. You'll see.

P: Oh, your changing the world all right, just like we did. But when you're 103, you'll back and seem how crude it all was. The good old ways were very good in many ways, but it's not until a technology becomes fully mature that it really is worth the trouble. Until then, it's more frustration than it's worth. But when that maturity occurs, the technology stops being high tech. Then something else – crude, frustrating, but with enormous promise – takes over and we forget benefits of the older technology. The automobile, the telephone, the electric light were like that in my day. Everybody was talking about the Internet. Maybe it's just an old man talking, but the net is nothing special to us today. The Gene Box is what everyone is talking about. But that will become mature someday, too.

C: Is that why you called it the waves of creative destruction?

P: Well, I didn't call it that. I borrowed the phrase. But, yes, every wave comes crashing over us – destroying the old ways but creating new, often better ones. It's hard making the transition, but that's how the world changes. Write that in your project.

C: Yeah, I'll create an animate that has a wave crashing on the net, moving it to the background and growing new genes in the foreground. Is that the point?

P: Yeah, that's the point, Craig. Good luck....

17

Why the Knowledge Revolution Needs a Cultural Revolution

Pita Enriquez Harris

There's something badly wrong with the human species when it takes us only four hundred years to exhaust our ability to perform a vital activity – one upon which all indications are we will need more rather than less in times to come.

The nature of the global work force is undergoing a significant shift in emphasis – away from traditional manufacturing roles and towards 'knowledge working'. As we enter the new millennium, it has been estimated that less than 10% of the world's workers are manual or factory workers. The 'Knowledge Revolution' is a third Industrial Revolution and the fourth 'Information Revolution'. Yet at the same time, we are hearing that those at the vanguard of the new Knowledge Economy are already tiring, slowing, stressed out by the weight of the information bombarding them, the information which has become the single most important raw material of their produce – knowledge.

Four hundred years ago the world experienced an event that led inevitably to a complete revolution in the way people communicated. The printing press altered something very profound about the way individuals dealt with written information – it encouraged them to read silently and to themselves. Until books were mass-produced, most communities would consider themselves lucky to have a reader amongst their number. Such a community would be able to become educated, maybe gain access to a book, which would be read aloud by the readers in the community to everyone else. Information was passed through oral traditions, before the invention of writing, to a large extent after the invention of writing, right up to the Gutenberg Age. And then, what a difference – information entered most people's mental processes via the eyes, we read to ourselves before passing on our acquired knowledge, often in writing.

People talk about the huge paradigm shift between absorbing information from a page versus a computer screen. The difference is trivial compared to the shift 400 years ago. By now we've already trained our minds to process information via the eyes – back then, it was a question of learning a *whole new way to learn*.

In 1996, extensive media coverage of a report commissioned by Reuters[1] to examine the impact of information overload upon industry executives ensured that 'information overload', a term already familiar to people in the IT industry, finally received some official recognition within the business community. At the 1997 World Economic Forum's annual meeting in Davos[2], countries' leaders were advised to 'go on an information diet'.

The team commissioned by Reuters went back to their studies one year later and produced *Out of the Abyss: Surviving the Information Age*[3]. It seems that the panic of the previous year was unfounded – we aren't yet actually dying from information overload; in fact managers are, on the whole less worried at the end of 1998 than they were in 1997. However, the findings still highlight serious problems in the developing as well as the developed world, with roughly half of all managers surveyed regarding information overload as a serious problem. True, information filtering and management systems are emerging that may be able to provide some sort of technical solution. However, as some knowledge management experts have been insisting, the best use of a company's knowledge resources is made where cultural change is the byword. It is simply not enough to install expensive information management and data mining software.

Those of us at the cutting edge of the 'Knowledge Economy' exist along a continuum of human information processing and knowledge production. It starts somewhere back in those days of the first printed books and reaches right into the future, such a future as exists now only in the realm of science fiction, when the written word has once again become unnecessary and communication is by pure thought. As we move into the knowledge era, we need to answer certain questions about the human brain's capacity to handle information.

What is it about too much information that overloads us? As an ex-cell biologist I ask myself biological questions about this – what is it about the human brain that makes it inefficient at processing written information? Is there some physiological reason and if so, can we ever overcome or out-evolve our physiological limitations? How much better can we ever hope to be at the job of producing knowledge from information? Or will we have to leave it all to technology to solve the problem for us?

'Our brains aren't wired to "multitask" the way our computers are,' says psychologist Larry Rosen, a human-computer dynamics expert and psychology professor at California State University–Dominguez Hills. 'We're taxing the limits of our human abilities.'

Professor Geert Hofstede, the founder and chairman of Netherlands' Institute for Research and Intercultural Cooperation, and celebrated organisational behaviourist, agrees that we have already surpassed the human capacity to process information. 'The human brain can only absorb so much information,' he said. In an article in the *Sydney Morning*

Herald, Hofstede is quoted as commenting that comparatively few resources have been 'put into educating people on how to cope with the
mass of information they are now confronted with.'

The whole question of timing concerns me. *Homo sapiens* has been around
for 200,000 years. That means that a newborn baby is, as far as we are
aware, just as well adapted to living in the Information Age as was a
newborn baby of the Stone Age. It would be a feeble species that did not
have some flexibility built in at some level – a greater ability to run fast
than was likely to be used on a daily basis; a greater ability to resist
disease than was absolutely necessary to allow us to reach reproductive
age, for example. A greater ability to process written information, in a
world where the written world did not yet exist is another clear example
of the flexibility possessed by our own species. We evolved to adapt to
challenges phenomenally different to those provided by our present society. And now that the livelihood of 90% of the world's population relies
on the knowledge economy we have to ask ourselves – have we come to
the limits of our evolved ability?

Not according to Professor Gabby Dover, Chair in Evolutionary Genetics
at Leicester University: 'It is likely that our ability to handle written information is an "exaptation" – a new function which arose as a by-product
of other, necessary adaptations. The human brain's potential to overcome
chaos is enormous – without the need for further change. There is enormous unused capacity as well as the people out there to solve the problem.
We no longer need biological evolution. Cultural evolution is much faster.'

Professor Dover believes that information overload is not an indication of
approaching the brain's limited ability to process information. Rather, it
is a problem caused by inadequate presentation of information by a
workforce increasingly measured by its written output. The message is
being clogged by misuse and abuse of the medium.

'Most of the information we come across is unimportant, jargonised or
duplicated. More bluntness is required. When Bob Geldof raised money
for Live Aid he said simply: "Give us your money. Now!" And the money
rolled in.'

For a future without more 'information overload', Professor Dover even
predicts the possibility of draconian measures, as people walk away from
the problem: 'You'll see events like a "National Day for Turning Off Machines". Or a week where no one can use the phone. It might be dictatorial
but it *is* a cultural solution.'

You don't even have to look back one generation to see the huge shift to a
knowledge economy. In real terms and for most working people, the advent of email and the World Wide Web is the crucial factor that has exposed
them to the abundance of information and the evils of information overload. But students aren't obviously complaining – they love the easy

access to study materials, the fun and informality of email and the quick fix and sometimes-sexual buzz of the chatrooms. Could it be a generation thing? You have to wonder about that Reuters report – how old are the 'executive managers' who form the core of their respondents actually likely to be?

My friend Jane and I share a post-school-run coffee. Jane is a mother of four and self-taught Internet whizz. She agrees that there may be differences to the way the over-40s feel about being bombarded with information. 'The post-war generation weren't taught to make decisions – we didn't have choices like children have today. The biggest part of coping with huge amounts of information is being able to make decisions about what to ignore. If you're insecure about being wrong then you can't make a decision.'

Handling information overload is as much about decision-making as about correct use of email programs, newswire services and filtering software. Noreen Mac Morrow, a Senior Lecturer at Strathclyde Business School delivered a paper at Online Information 98, which intrigued me by daring to say what many information industry pundits never would: advising staying incommunicado as a strategy for dealing with information. It was a similar message to that which I had read in an interview with Umberto Eco ('A Conversation on Information'[4]): 'At my level of "visibility", my problem is to avoid the message ... otherwise I will be destroyed by the number of messages. My problem is not to answer the telephone; my problem is to destroy the fax, the unrequested fax as soon as it arrives.'

Noreen Mac Morrow agrees that information overload is as much a cultural as technological issue. 'We gather more and more information but allow ourselves less and less time to actually absorb it ... part of the problem is finding that reflective time to be able to out the pieces together in a way that is meaningful.'

According to Noreen, information literacy is the key. 'This is more about the culture of use – retrieval has become much, much easier. Instead, we need to focus on what people do when they have the information. For all the talk about "user requirements", we're still very fuzzy about the way people "in use" value information.'

The point is that retrieval, as smart as it can get, can never truly relieve us of information overload. Think about it – how smart could it get? Could it get better than if you yourself did the final filtering? And it is a rare manager who ever finds *just* enough information, even for himself or herself. That is what we use filing cupboards and hard drives for – to file away pieces of information 'just in case'.

Noreen again: 'People say that they want "everything" relevant – to cover themselves. It is the security blanket of having 5 million papers and saying "I've exhausted all possible research".' We need a whole new way to think about information retrieval; an attitude of 'just in time', not 'just in case'.

When the human brain sifts through information, it tries to make sense of it in the manner to which we have been biologically programmed to communicate – it tries to turn it into a story. This is why you feel miserable and tense after a long day of gazing at a spreadsheet. This is why most people aren't capable of making a living as a physicist or mathematician. It takes a particular type of mind to make stories out of the totally abstract language of mathematics.

Dr. Michael Stein, a Commissioning Editor with Blackwell Science, considers the issues faced by academic publishers, trying to produce concise information packages (books and journals) in an environment of mounting quantities of information. 'The amount of medical literature is doubling every seven years, according to MEDLINE,' he tells me. 'The problem with information overload is that people are unable to make a coherent story out of it. They try to bring in all this disparate information but what really makes a good story, or a good textbook, is a distillation of wisdom. All our best teachers have the story-telling skill.'

If the transfer of the written word from the page to the screen is relatively non-revolutionary, the invention of hypertext *is*. It introduces a whole new perspective to the story-telling paradigm of human communication – the story that is bifurcating, labyrinthine, and always unique.

Michael Stein reflects on the implications of this. 'The problem with the Internet is all the amazing amounts of information. Certain people have the ability to navigate through that and create their own story. But most people aren't actually that creative. They want to be told, they want to hear stories.'

Hollywood understood this principle from the beginning, as did Henry Luce, the founder of Time Inc., who responded to the criticism that his magazines relied too much on anecdotes, thus: 'We didn't invent the idea of delivering news through stories about people. The Bible invented it.'[5]

Now, even Big Business seems to be embracing this idea. In his paper to the 1999 Knowledge Management Conference and Exhibition (held in London, March 1999), David Snowden of IBM Global Services spoke about a new knowledge management (KM) practice of collecting and storing the kind of anecdotes about the business and using this database of stories to the advantage of the company. If adopted generally as a 'KM Technique', this will represent a realistic, duplicable approach to the problem of how to capitalise on the tacit knowledge within an organisation. As Thomas Stewart writes in an article for *Fortune,* [5] 'Nothing serves a leader better than a knack for narrative. Stories anoint role models, impart values, and show how to execute indescribably complex tasks.'

In the short term at least, we cannot escape our essential biological nature. In current widely practiced management structures, decision-making is commonly avoided to limit acceptance of responsibility. Under these circumstances, a tension exists between the urge to gather information and an inability to cope with too much, between the abundance of fragmentary information and the need to synthesize stories from such lore.

The strength of our species is that we are organic learning machines – there is scope for a certain amount of rewiring of our mental processes. It has been done before, when we made the leap from the oral to the written tradition, from the handwritten to the print. Another such leap is possible, but the task requires nothing less than recognition of the fact that much of the old rulebook must be thrown out and new skills of information literacy must be learned. And we need the software tools to help, acknowledging that while we can be greatly assisted in the retrieval, filtering and sorting of information, assigning value judgments and decision-making requires some processes which remain mysterious, intangible, and dependent on constructions of our own society. The success of the 'Knowledge Economy' depends on a cultural revolution more than it depends on IT.

References

1. Lewis, David (1996). *Dying for Information: An Investigation into the Effects of Information Overload Worldwide.* Available at http://www.reuters.com/rbb/research/overloadframe.htm

2. 'Stay on "info diet", technology experts warn business.' Reuters/Wired press release, February 2 1998

3. *Out of the Abyss: Surviving the Information Age.* Available at http://www.reuters.com/rbb/research/newresframe.htm

4. Coppock, Patrick (1995) in '"A Conversation on Information", an interview with Umberto Eco by Patrick Coppock, February 1995,' in *Multimedia World* http://www.cudenver.edu/~mryder/itc_data/eco/eco.html

5. Stewart, Thomas (1998). 'The Cunning Plots of Leadership' in *Fortune,* September. Available at: http://www.pathfinder.com/fortune/1998/980907/lea.html

18

The Evolution of Media Librarians: Charting the Future

Barbara P. Semonche

Introduction

Forecasting the future is a heady task. I am reminded of how Yogi Berra, former New York Yankee catcher turned pundit and raconteur, responded when asked a question by a reporter about forecasting his team's future: 'Prediction is very hard, especially when it's about the future.'

Berra could take comfort in a book that documented well the capricious nature of forecasting. *Today Then: America's Best Minds Look 100 Years into the Future on the Occasion of the 1893 World's Columbian Exposition*, published in 1993, offers ample evidence of the folly of predictions.

In the early 1890s, a news agency commissioned 74 prominent Americans to write brief essays on what life would be like in 1993. Newspapers published essays as part of the fanfare for the future-oriented Worlds' Columbian Exposition, which opened in Chicago in 1883. For the most part, the forecasts have turned out to be not just wrong, but hilariously wrong. Examples: journalist Walter Wellman correctly forecast the coming of the airplane but he thought it would be powered by electric batteries. Taking transportation one step backward, Wellman saw no future for subway trains, suggesting that Americans would travel instead in elevated trains. Not entirely wrong, but then again not exactly on target. Curiously, none of the forecasters predicted the arrival of automobiles. A few considered the possibility of telephones and telegraph, but predicted that transcontinental mail would be transmitted in pneumatic tubes. Correct predictions included the arrival of an income tax, air conditioners and the vote for women. Why do forecasts err? Firstly, significant changes occur constantly all around the world, but we are totally unaware of – or unimpressed by – most of them. Horseless carriages were already in use in Europe before 1893, but American forecasters were either unaware of them or unimpressed by them. At best they saw no future in them.

Secondly, recent events, especially those experienced personally, dominate our thinking about the future. Railroads were developing feverishly in the 1880s and 1890s and it took little imagination to think that they would become faster and more widespread in the future. Despite my best efforts, there is a good chance that my vision of information products and services in the twenty-first century will be biased.

At the end of the twentieth century, the Internet, the World Wide Web and e-commerce dominate the news, the culture and the technology of our lives. They are fast becoming ubiquitous. It is too easy to forecast their continued, even accelerated pace. But, even forecasting here is shortsighted. Here's an example: early in 1998 forecasters said business-to-business e-commerce might rise to $300 billion in the United States by 2002. Not so, says the U.S. Department of Commerce; business-to-business e-commerce will be more like $1.3 trillion in 2003.

Predictions about retail web commerce have missed the target also. Last year, forecasters said this sector would amount to $7 billion by 2000. By 1999, the U.S. government reported that the transactions probably totalled twice that in 1998 and could reach $80 billion by 2002.

It does not require a strong imagination to make the connection between information and e-commerce once the realisation is made that we have left the information age and are embarking into the era of information as a commodity with intrinsic value.

My specific task is present my vision of the future of information for news librarians, the organisations they work for and the journalists with whom they work.

The Past as Prologue

News librarians (now referred to more often as 'news researchers') have evolved from early roles in the pre-digital age. As 'morgue clerks' they were kept busy warehousing obituaries, metal engravings and glass negatives; later, as 'information gatekeepers' they controlled the ebb and flow of news clipping files and photos; then as 'information intermediaries' they wielded power over journalists as database retrieval experts; and now in the last decade of the twentieth century, they have morphed into overseers of all departments' potential for research, repackaging information and for discovering different information delivery mechanisms.

On the eve of the twenty-first century, news library leaders are forming alliances with other departments such as advertising, marketing and networking. They are coordinating aspects of research and data retrieval into new production system designs insuring the needs of researching and retrieval are recognised and incorporated in the proposal stages, thus preventing costly mistakes when systems which do not provide the full range of capabilities are selected.

Building on the traditional skills they have always brought to the function of information management, news researchers are demonstrating their abilities to refine and redefine their newsroom roles into far-reaching corporate information officers. In the process they are discovering that they do not work for a single medium anymore.

These late twentieth century news researchers are becoming knowledge managers recognising the multimedia intellectual assets of their organisations and seeking ways to capitalise on them. In doing so, they have left the information age and are entering the information economy era.

News librarians have a long history of serving two primary functions in their media organisations: as archivists of ever-growing, increasingly large, unwieldy, complex multimedia materials and as researchers for data on deadline and in-depth investigative reports. There are obvious pressures to maintain quality archives for several types of clients: journalists, media managers, network technologists and the public. Simply said, news librarians are big-time producers of information as well as information consumers. And it does not take a leap of faith to predict that these major functions will continue into the early decades of the millennium.

Issues

It would be an injustice to project a bright, shining future for news librarians with few problems that technology could not solve. The media knowledge managers of the future will inherit the challenges those of us in the past have barely defined.

Mass communication is in general decline. Major metropolitan newspapers and network broadcast stations are losing market share. Competition for readers, viewers and listeners is coming from new media as Internet and the World Wide Web. Inroads to market dominance is coming from dizzying array of 'niche' publications such as special interest magazines and newsletters, upstart alternative or 'underground' press and 'street' publications. Information overload, as well as accuracy, quality and fairness of information are challenges that will never disappear regardless of the power of technology or the talent of news library managers. Newspapers have long been champions of public access to government documents, but with the increasing complexity of many digital public records containing private data, the ethical and privacy concerns are wide and deep. News librarian roles in the acquisition and storage of this information have yet to be explored. Discovering the questions to ask, then ordering their priority in the realm of the industry will be every bit as critical to successful media management as coming up with the answers.

Major newspapers started closing or merging operations in the late 1950s
as television news broadcasts became the 'medium of choice' for most
Americans. Most afternoon daily newspapers ceased publication in the
1970s. There is every reason to believe that this trend of media consolida-
tion will continue further reducing employment opportunities for news
librarians. In addition, a little known problem will continue to face the
next generation of news librarians. That is what to do with the increasing
gargantuan loads of backfiles (daily photos, news clippings, reference
materials and electronic data) that will become the residue of these de-
funct dailies. Much of this material has already been lost, carted off to
trash heaps, but what of the remaining? Who owns this information?
What is the value of this information? What are the options to archive this
material? Can it be repackaged at reasonable cost? What is the potential
market? And finally, what are the consequences of not archiving this
information? News librarians in the next generation will face these and
other questions as they discover how best to serve their company, their
profession and the entire information industry.

Information Ethics: News Librarian Style

The problems of information ethics centre on how librarians can balance
the interests of their news organisations with the privacy rights of the
public clients. For example, subscribers to full-text electronic news
databases and as well as public citizens who solicit special research directly from the news library staff. Passing on intriguing research inquiries
from public to the news staff might legitimately be unethical. While this is
not a concern to some clients, others might legitimately feel otherwise. A
delicate balancing act between the needs of the clients and the needs of
the newsroom begins. The ethics of information selling concern everyone.
Defining public access to information, determining the value and accuracy of information, and establishing its cost has been a part of the news
librarians' past agenda; it will likely continue into the future with even
greater complications looming. Some news libraries have simply opted
out of this customised 'research for sale' because of the ethical conflicts.
What will future news organisations expect from their 'research center?'
Profit, surely, but ethical issues about information use or misuse may be
an unwelcome by-product.

There are related critical ethical issues involved in selling, repackaging
or re-purposing data. For one thing, who will be liable when incorrect,
incomplete, or out-of-date information is sold? Who will be responsible
for insuring the quality and accuracy of information in its myriad formats? These are certainly not easy questions; even more important they
do not have easy answers, but it is a certainty that they will be part of the
twenty-first century's dialogue on information policies.

In addition, there are potential ethical conflicts when news librarians find themselves searching public data (to which they are legitimately entitled) and discovering private information about individuals embedded in it. These examples surface when tapes of raw data from public agencies are acquired and stored for purposes of computer-assisted investigative reporting or CAR. Solutions may involve devising programs to 'mask' or delete private, personal information fromthese records. As CAR programs increase in media organisations, these concerns will multiply.

Database Quality Control

Daily newspapers, hourly broadcast programs and weekly news magazines all are committed to accuracy and fairness. Yet mistakes are made and while usually corrected they are not always updated in electronic databases. The pressures of time, limitations of staff and the huge amount of information flooding the market makes this challenge the most difficult to address. No one expects these problems to disappear anytime soon. Perhaps technology will create better solutions, but probably not in the foreseeable future. Whatever the solution, it will definitely require the collaboration of talented, dedicated professionals in at least two domains: informationscientists and systems engineers.

Reference Material Evaluation

Selecting reliable, credible information in about current reference resource material is not unique to news librarians. However, comparison about these products and services have eased and accelerated since the advent of the electronic discussion list, NewsLib (http://metalabunc.edu/journalism/newslib.html). Recommendations from veteran news librarians on what CD-ROMs, reference books, web links and online databases are most valued are displayed at this URL: http://metalab.unc.edu/slanews/. News librarians looking for fair and honest assessment of what works, what lies ahead and who knows can find answers there. The next generation of news librarians will have the benefit of these sources and doubtless contribute new and updated material as time and information needs change. Without electronic discussion lists and professional web sites the next generation of news librarians would be hard pressed to reliably uncover the good, the bad and the ugly stories about information quality.

Trends

The trends in the news media industry seem to continue the reliance less upon production staff and more on editorial staff. Technology is the driving force here. The explosion of the information age has special meaning for news libraries. It became a sort of 'Goodbye, Gutenberg; hello, Mr. Chips' scene.

In the late 1970s and early 1980s, many newspapers went from hot type to cold type, that is, to electronic front-end typesetting. Even smaller newspapers could amortise multi-million-dollar computerised typesetting investments in a relatively short period of time. Sadly, this technological advancement eliminated long-time newspaper craftsmen, typesetters. But, as a result, within a few years, newspapers were able to reduce the number of their employees significantly with a substantive improvement in productivity. This extraordinary technological development meant dollars saved and profits increased. Some have speculated that newspaper libraries were able to reap some of the benefit of this increased corporate savings because that era became the benchmark for news library automation.

In the latter decades of the twentieth century, more sophisticated information services and equipment arrived. Professionally trained, experienced information specialists figured more prominently in news library staffs in the 1980s and 1990s. This trend was, however, just another step in the evolution of news librarians, that of developing a full partnership with journalists and the entire publishing enterprise.

Broadcast and news magazine librarians joined their newspaper comrades in tackling the technological challenges of the latter decades of the twentieth century. Advanced technology such as computer-output microfiche, computerised indexing, automated full-text retrieval systems, digital photo archiving, standards for electronic publishing and archiving, metadata, online database searching, statistical analysis data management packages, computer-assisted reporting and research, geographic information systems, media polling, and the Internet with its graphical browsers appeared on the scene to enhance and complicate media research, storage and delivery. No sooner had one archival technology been selected and implemented than it became obsolete and the legacy files needed to be converted to newer information storage systems. Still, news librarians were readying themselves with the professional skills and talent to select, manage and master such electronic wizardry.

Predictions about newspapers disappearing in the twenty-first century are silly. Undoubtedly there will be fewer dailies, but newspapers (including the community or non-daily newspapers) are likely to continue being a reliable, timely, cost-effective, if not exclusive, delivery mechanism for news and advertising for quite some time.

Some predict that there will be no central newsroom in the twenty-first century resulting in no 'final' edition. This is a revolutionary prediction but possible. However, there are questions. What will happen to the concept of the 'paper of record?' What will be the 'archived' edition? Furthermore, does this 'powershift' mean that there will be a blurring of news staff roles? Will print and broadcast reporters still have a role? Or will there be 'bionic' journalists racing to news events equipped with suite of high-tech portable devices (lightweight yet powerful laptops, digital video cameras, digital assistants, colour scanner, printer, cellular modem connections, continuous speech-voice recognition system, global positioning system, television receivers, digital video disk drive and more) all capable of producing news for print, online and broadcast operations? And if those are the tools of the twenty-first century journalist, what are the ones for the twenty-first century news librarian cum knowledge manager? Mirror-tools of the journalists'?

There is an undeniable close relationship between journalists and news researchers. Will news researchers become 'portable,' too? Perhaps we already have with our intranets and virtual libraries.

Visions

It is the dream of news librarians everywhere that if we could just get all the in-house news information and data we have from all the current and past formats (print, microfilm, digital, audio, visual) into one accurate, attractively designed, intuitively searchable, quickly and remotely accessible, reasonably priced, undeniably profitable, reliably constructed, conveniently updatable, easily compressed, confidently secured, potentially convertible and highly respected database, then the information world would be at peace and librarians would rule, benignly, of course. When this happens it will signal that the highest level of information evolution has been reached.

Such a Herculean task will be difficult, if not impossible. It is better left to the next generation of news library wizards. This next generation of news librarians (whatever they will be called) are likely to become the heirs to some of the most serious archiving problems ever identified. Those full-text news archives which debuted in the late 1970s and early 1980s went through several 'conversion' formats. However, unifying them into one cohesive, easily accessible, confidently accurate, current and comprehensive database in still not a reality. Add to that the archiving challenge, the multimedia demands of digitised photos, audio and video, not to mention the diverse contents of web products and the dream of an integrated archiving systems remains elusive. This is a mighty challenge and one that news researchers, information technologists, networking administrators and database designers of the future must solve.

However, archiving is not the only concern of this or the next generation of news librarians. The never-ending challenge of providing the most current, comprehensive and accurate information on the widest possible number of topics on deadline to news staffs will continue and doubtless expand. News librarians are already part of special teams assigned to serious, in-depth investigative reports on complicated issues and trends. It is easy to predict that this level of research will continue with some modifications, namely that news librarians will expand their roles as cyberscouts, coaches and trainers for journalists undertaking conventional and electronic research. These and other roles news librarians will play in their media organisations as they become knowledge managers of news information assets and corporate leaders in new, and profitable information products.

The following is a list the author developed after speculating on the potential job titles and leading roles news librarians are likely to play in the media industry of the future. There is definitely room for expansion and redefinition.

- Research Stars
- News Research Editors
- Research Training Coaches
- Database Creators
- Integrated Archive Specialists
- News Database License Counselors
- Database Quality Control Experts
- Intranet Architects
- Web Content Developers
- Network Wizards
- Technology Scouts
- New Media Product Designers
- News Product Marketers
- Information Product Evaluators
- Ethics Watchdogs
- Corporate Information Officers
- Corporate Knowledge Asset Managers
- Knowledge Strategists
- Knowledge Editors
- Expert Communicators

For a model of future news librarians, read on. Anders Gyllenhaal, executive editor of *The News & Observer* in Raleigh, North Carolina, published in January 17, 1999 this announcement in a column on the newspaper's plans for the new year:

'For the past dozen years, Teresa Leonard has quietly had her hand in just about everything that goes into *The News & Observer*. She's the director of news research, which means her 15 staff members handle the archives, databases and founts of information the paper is build upon.

Starting this month, Teresa has taken on a new role as well.

She'll loan her considerable research skills to a deep review of how the region's growth should guide the way the paper itself is assembled. Teresa and a group of others will spend six months analysing the changing shape of the Triangle with questions such as these in mind: How can we better serve the fast-growing parts of the region? What should the territory for the various editions be? What's the best way to organise and present our news reports for the different parts of the Triangle?

This will be one of the major undertakings in a year in which we hope to make a number of improvements to *The N&O*.'

Obviously, this kind of trust in the ability of a news librarian, albeit a gifted, dedicated and experienced one, to play a leading role in the organisation's future is extraordinary. It demonstrates clearly what opportunities for knowledge leadership may lie ahead for the next generation of talented news librarians.

Predictions

The trends, issues, currents and cross-currents in information technology, library education and the media defy facile identification or a complete inventory. This is where one's individual biases and limitations affect the outcome. Having said that, I will venture a list, in no particular thematic or chronological order, of my predictions for the future. The more well-informed among us will recognise that some of these predictions already have their antecedents in the later decades of the twentieth century.

1. Information products and services are no longer the singular domain of librarians and researchers. New partnerships between other corporate news departments, specifically data systems specialists and technicians, will accelerate the development and marketing of new media products.

2. Research will not be the singular province of librarians. Journalists and librarians will partner in a widening variety and complexity of research tasks. Computer-assisted research and reporting (sometimes referred to as CAR-R or simply CAR) will accelerate 'team' journalism. Team members include not only the investigative re-

porters special project editors and photojournalists, but statistical analysts, database managers and news researchers as well. News researchers will continue to receive credit lines for their contributions, an innovation in the 1990s. Examples of these types of award-winning projects and the contributions that news library researchers made can be found at this URL: http://metalab.unc.edu/journalism/awnwslib.html

3. More news libraries will face closings and mergers resulting not only in unemployment but in challenges about how to store and share large backfiles of text and images. Conversions from one storage medium to another will increasingly trouble news archive integration, access and quality.

4. More independent or entrepreneurial knowledge workers will emerge in the news industry. News organisations will expand outsourcing of their information staffing needs. 'Itinerant' librarians (aka cyberscouts, 'gumshoe' librarians, independent data contractors, information brokers, communication consultants) may become the new 'hired guns' on the media information frontier. These 'information specialty itinerants' will actually travel or telecommute to a number of their media clients. Conflicts of interest are a definite possibility as well as challenges to data security.

5. Career shifts for news librarians will increase between private, public, government and academic institutions. News librarians should be on the lookout for new employment possibilities in academic posts, as sales representatives with software vendors, as independent consultants or as telecommuters. For active learners with demonstrated skills in marketing, product development, team building, opportunities for advancement in the main corporate hierarchy exists, but it means leaving the library. Knowledge managers with MLS and MBA degrees will be highly sought after.

6. Some of the most intriguing knowledge jobs, however, can best be described as knowledge reporter/editor. Even if news librarians have the skills to frame and structure knowledge, few have the time to put what they know into a system. People who have demonstrated skills in extracting knowledge from those who have it, who are then able to reorder it to a form anyone can use, and can effectively, efficiently update and edit that knowledge will be prized. These abilities go beyond 'building data about data' that is traditional among librarians. These skills are not really taught anywhere, but the closest approximation may be found in journalism schools. The next generation of news librarians with journalism and information science will have the best opportunities to become the high-tech communicators of the future.

7. Journalism and mass communication schools will continue their mergers or consolidations with schools of information and library science. Joint degree programs will be launched supplying media organisations in the future with multi-talented, diversely-trained, ethically-focused news staff.

8. Distance learning programs will dominate the library education field for the non-graduate degree student. Undergraduate programs in information technology will increase, supplying competition for entry-level news library positions. As these neophytes proliferate, greater reliance on 'continuing education' programs will rise. Electronic mentors, professional chat rooms, desk-top delivery of computer instruction, regional workshops will expand. Professional certification will gain credibility as top management desire documented value for their investment in information specialists.

9. Media library workers who are not advancing within the corporate culture will be ripe for organisations touting 'collective bargaining' and job security.

10. Mid-career journalists and news librarians will return to college to jump-start stalled careers.

Further, newspaper librarians will need to prepare themselves for evolving, expanding news roles. The effort will require not only new skills and knowledge, but a willingness to actively participate in the development of new technology. Times and methods have changed, and so has the media industry and its libraries. Regardless of whether the tools of the twenty-first century news librarian remain scissors, paste pots, ledgers, file cabinets and telephones (as a few almost certainly will), or become computers, imaging archive systems and a host of other cyber age technology, the vital ingredient will always be the individual's professionalism and partnership talents.

In a very short time librarians 'morphed' from 'morgue' clerks into database designers, web architects, knowledge engineers, internet coaches, authors, marketers, product developers, copyright specialists, contract negotiators, and independent information professionals. Still, there continue to be elements with are common to all news librarians over time and distance. In describing these common professional traits, one astute news librarian observed: 'Our responsibilities are to discover, nurture, cultivate information; harvest it, keep it clean, store it, protect it, and share it.'

That seems good advice for news librarians then, now and in the future.

19

Information Ontologies for a Digital World

Jonathan Raper

Introduction

Before we know the future of information we must know what information actually is. Yet even the definitions of information we have seem to be changing as we move into a digital world. In order to tackle this information conundrum I have turned to ontology in this essay for a theoretical definition of information that is independent of the medium in which is recorded. This approach delivers a stable platform from which to observe information in a time of profound change. Once we know where we are then we can try to see where we are going.

The study of ontology is the process by which phenomena are defined and given significance; essentially, an 'ontology' is a 'specification of a conceptualisation'. In philosophy ontology has been studied since Ancient Greek times by exploring possible categorisations of phenomena. Following Plato, 'metaphysical realists' argue that everything can be divided into 'particulars' (individual things), e.g. a mountain, and 'universals' (repeatable phenomena). Universals can be further defined by their 'properties' (such as shape), the 'kinds' of which they are members (e.g. taxonomic group) and the 'relations' (e.g. topological) they enter into. As a consequence every 'particular' can be said to exemplify a set of 'universals', e.g. there is a mountain which is flat topped, in the Pennines and at the edge of a range (Ingleborough, for the curious!). On a philosophical level, the discovery of repeatable 'universals' has been taken to imply (by realists) the existence of repeatable structure in the world. This then allows the construction of ontologies with a direct correspondence to reality.

In the 20th century some philosophers have added 'propositions' to the ontology of particulars and universals in order to encompass the abstract things we create by thinking, speaking and acting. These abstract things may be concrete with an existence at a specific place and time or they may be imaginary and 'intentional'. Propositions can be defined as language-independent assertions about these abstract things, which can be either true or false. For those philosophers who can't accept that there are true propositions about things that don't exist, it can be argued more simply that propositions are 'properties' of the more generic 'states of affairs', whether real or imaginary ones.

These ontological debates give us a theoretical apparatus by which we can formally conceptualise information. Accordingly, we can propose the conceptualisation that 'when properties of "states of affairs" are represented and communicated in any natural or symbolic language, then the result is information'. This allows information to be a very basic process and separates its essential ontological nature from any impact that the communication of it may have. On this definition information needs no mind to create it: information signals are generated, communicated and received in nature, e.g. human DNA from parents to child. Information representations can, therefore, take many forms.

Human cognitive systems are, though, needed to sample and order information, for example, into scientific taxonomies, library indexes or collections of observations. In this sense, an 'information ontology' is a commitment to use a specific vocabulary when representing and communicating information, in a way that is consistent with the underlying ordering. Human cognitive systems are also used to invest information with meaning to create knowledge, which is the 'justified, true belief' of the individual. Scientific studies of the world have involved centuries of collection and communication of information and its synthesis into knowledge, which is continuously tested and revised by researchers.

In a pre-digital world analogue paper-based collection and communication of information representations placed practical limits on the amount of information that was routinely collected and ordered. In the digital world the cost of collecting and communicating information representations has fallen dramatically and the amount of information collected (by sensors) ordered (by computer programs) and used by people has rocketed. As a consequence the information ontologies of the analogue era have begun to break down. No longer must information be defined by the medium in which it is encoded.

As symbolically encoded information representations in digital form have new properties, so have new ontologies for information become available. How then will information be recorded and ordered in a digital world? How will information be selected and signified and who will control these processes? Will these new information representations change the way we think about the world? The answers to these questions seem likely to play a critical role in the development of information in the next millennium and are explored below. Speculating about these questions is the way I have chosen to look into the information future.

Recording and Ordering

The recording of information in digital form creates new types of owner-ship of intellectual property. The ability to gather and collate a vast number of facts using computers has already forced a reconsideration of the mean-ing of 'compilation' in the context of copyright law. Even if the facts require no skill to collect and the compilation involves no originality of arrange-ment, the new European Union 'Database' Directive (96/9/EC) protects databases from the extraction of information without the creators' per-mission. Hence, the time taken to collect and arrange facts has created a new form of digital information possession. However, the usual consid-erations used to define the copyright in a work will no longer be adequate for a digital database. Such databases may never be finished in any de-finitive sense (e.g. when a database is updated every second); and, there may be no natural owner (the contents may be accumulated by recording the operation of many sensors). I foresee a battle in the next decades about the appropriate definitions to use for closure and creator in digital infor-mation possessions.

Most digital information is currently encoded in a small number of widely readable formats created by the major operating systems and applica-tions. Sending an information representation through the Internet under these conditions is equivalent to sending a work through the post with-out wrapping. Users rightly fear that the exclusive ownership of their information is placed at risk in a digital world: anyone can read or copy their information without leaving any trace. The natural solution to this problem is for users to employ secure coding and communication meth-ods such as Public Key Cryptography to encrypt their information. Now that governments appear to have accepted that the availability of strong encryption is a civil right and are no longer insisting that the public give up their private keys to 'trusted third parties', I can foresee a world in which encryption plays an important part. Without encryption informa-tion ownership can only be relative and not absolute. We can of course expect tragedies like the accidental loss of one's own private keys and the need for the police to make de-encryption ambushes on cyber criminals!

Information orderings in the form of metadata schemes will play a crucial part in the digital world. Metadata schemes must take a semantic view of information in order to specify the attributes and their associated do-mains to be used in information representations. The term 'ontology' has been used in computer science in a narrow sense to specify such a seman-tic view. In the wider sense of the term used here an information ontology is the superset of all the semantic, coding, vocabulary and definitional issues associated with information. I would suggest that the developers of metadata schemes must consider the wider sense of ontology to avoid partial solutions that get embedded in practice before all their implica-tions have been fully understood.

Currently there are metadata schemes in preparation in hundreds of economic sectors and government departments in dozens of countries, many of them overlapping. There will have to be a reduction in the number of such schemes for information interoperability and 'knowledge' management to become a reality. The creation of high-level information ontologies to provide an organising framework will be the key to this task as in the World Wide Web Consortium's Resource Description Framework (RDF) project.

Signification and Control

Since representations of information are the foundations for public debate, policy development and decision-making in developed societies, profound changes in the technological environment of information must also change the relations of power between groups in society. This is because information has become a key driver of social and business relations, creating new opportunities and making for new inclusions and exclusions in society. As in all such changes there are winners and losers in influence terms. Over the medium term the losers will include: the information gatekeepers who have traditionally controlled the flow of analogue documents and communications within organisations; publishers who no longer control the flow of publications to the market; and governments that seek to restrict the information flow within society.

The winners in influence terms are those who now control the new flows of information. These include computer operating system makers like Microsoft, the owners of data format 'franchises' like Adobe's Acrobat portable document format, the owners of digital communication networks like the cable companies, the Internet gatekeepers like the domain name registries and the new information 'portals' on the web. The latter include indexes of information on the web such as Yahoo! as well as search engines like Altavista. If the web remains the dominant organising metaphor for information delivery in the digital world then portals will have enormous influence as they will be both a marketplace and a library rolled into one. A key question will be what kinds of communities will support a portal in the future: will they be based on national states, language groups, service provision or online communities?

The only threat to the web as the pre-eminent infrastructure of the digital world is the opportunity provided by digital television delivered by satellite or cable. The 'set-top box' will have the capability to offer interactive services that are likely to grow progressively more sophisticated as the number of digital television subscribers' increases. This will give the media companies the opportunity to deliver information alongside entertainment, maintaining (and extending?) their influence in the digital world. The web versus digital television promises to be one of the key battlegrounds in the information future.

The selection and signification of information will not just be a question of the channel/format through which the information is conveyed. Another key question will be what kinds of business practices, economic models and legal regulations will emerge to dictate the nature of information commodification in the digital world? The flexibility of digital information means that the next generation of information providers can reinvent information products. New strategies that have already emerged include free sampling of the goods to lure consumers into bigger purchases, product release in instalments with payment through subscription, and disaggregation of information into small units to facilitate micro-payment pricing. Such strategies will define new marketplace(s) for digital information: in many cases they will create new information ontologies by default.

The Informational Society

The changes in the recording, ordering, signification and control of information in a digital world have the potential to impact society profoundly. This claim is not based on the kind of technological 'boosterism' which is frequently heard from some quarters. It is based on the clear evidence that developed societies must and will change to adapt to the new power relations and economic opportunities that are emerging in the digital world. The impacts of these changes are at the time of writing only apparent to a minority of the population in many societies, but the information future will be determined by what happens when the rest of the population is enmeshed in what Castells has called the 'Informational Society'.

What then is the role of the academic, the librarian and the information professional in this informational society? The answer is surely a determination to reproduce the principles of the past in the information future: equality of access to information, the preservation of the information commons, making a reality of the learning society and guarding against information totalitarianism. If we succeed, then the information future is ours too.

20

The Information Specialist as Fulcrum

Gary Marchionini

It is a safe claim that the twentieth century has been dominated by technology. The airplane, atomic bomb, and television are but three examples of transforming technologies of the first half of the century that changed the way we live and think about our world. Together with biotechnical advances, computer and communications networks, collectively known here as information technologies (IT), are causing similar transformations at the close of the century. Moore's law predicts that computational power doubles every 18 months – the impact of such change can perhaps be grasped if we take a biological example of cell growth in one's body reaching such rates! These rates of growth cannot be ignored and it seems obvious that information technology has driven many of the developments in the industrialized world for several years. In the midst of these changes, information specialists, those humans who intermediate human information needs and information resources have emerged as a new kind of navigator/guide in the emerging cyberspace. Unlike the engineers and technicians who invent and build the IT engines and codes, information specialists begin with the human need for information and leverage IT to serve those needs. In the past, this intermediation mainly linked humans to physical resources such as books or other people. Increasingly, information specialists facilitate linkages between people and IT systems, manage interactions among multiple humans through IT systems, intermediate between multiple IT systems, and coordinate the interactions among complex organisations and institutions with the aid of IT systems. In this short essay I argue that if trends in IT development continue in the decades hence, the information specialist will be increasingly called upon to balance the possibilities with the realities of human life – will become the crucial fulcrum upon which the balance of human interests with technical progress depends. This is a role that goes beyond the traditional role of assisting people to cope with the basic abstractions known as information.

Newton crisply stated the requirement for balance in the physical world in his third law of motion. We can easily see evidence for similar balance in the social world by observing intergenerational flux, political parties, and other dialectic balances. As our world becomes increasingly dependent on IT, we will surely see counterbalances in forms ranging from outright Ludditism to increased emphasis on humanistic and spiritual values. As the human and technical orders interact to define and redefine perspective and power in the physical world and gain control in the virtual/informational world, information specialists will find themselves in the thick of the fray. Stuart Kauffman and other complexity theorists argue that interactions among complex elements invariably lead to unpredictable emergent phenomena. The interactions of human information processes and needs with IT will surely lead to unexpected conditions, and information specialists must be both creative and tolerant to help their clients manage the information challenges sure to come – they must serve as the 'attractors' that insure homeostatic stability in complex systems. Information specialists must be creative in dealing with the new possibilities that emerge from the interactions of libraries and IT and the broader interactions of people and IT. Just as importantly, they must be tolerant of the multiple reactions people will inevitably have to new phenomena. By serving the diverse needs and reactions of the global population rather than narrow interest groups, information specialists will extend the basic trust that is so essential to social fulcrums. This dichotomy is of course drawn artificially sharply and there are already many cases where art and technology interact to form genres and functions that are truly emergent in nature. The point, however, is that the challenges to information management will become more difficult in the future, and information specialists will find their services in more demand than ever before to balance the need to obtain succinct and reliable information for timely decision making with the huge densities of relevant information not just available but bombarding us before we even know we need it. To illustrate these roles, I offer three trends – ubiquitous access, personalisation, and microcharges – that each provide new information management challenges as well as new bases for reaction and the consequent need for balance.

Ubiquitous Access

Mark Weiser articulated the trend toward cheap, portable, networked computers with the term 'ubiquitous computing'. As wireless networks become more viable, ubiquitous computing supports ubiquitous access to information resources and tools. Today's personal digital assistants and cell phones provide computational and communicational facilities to ordinary citizens wherever they are and whenever they desire such services. Active badges and jewelry or clothing with embedded IT will untether us from our desks and give us the IT power of office workstations in unobtrusive, on-demand fashion. As our personal, ubiquitous IT interacts with that of others, new possibilities arise for communication, collaboration, and exchange. Likewise, new potentials arise for abuse, and demands for secure information management follow.

Just as people will carry IT with them everywhere, the environment itself will be increasingly augmented with IT units – offices, automobiles, and homes monitor and 'act' based on signals from their owners or according to programmed rules. By combining sophisticated sensors, computational engines, and wireless communication, people can stay 'in touch' with their possessions across spatial boundaries – including their growing external memories packed with information acquired purposefully, casually, or automatically. Consider the information management issues we face with the files, bookmarks, and downloads we purposefully save on our workstations today and shuffle between home and office. Ubiquitous access wherever we happen to be can only complicate the organisation, retrieval, and display challenges even as the technical challenges disappear. Add to this image the mass of information that cheap memory, computing, and communication allows us to collect automatically for later 'mining' activities, and the information management challenge is qualitatively more complex. Whether these data streams result from processes ('agents') we launch with particular goals or automatic logs compiled by the countless IT components, we surely will need new strategies and assistance to deal with our information treasures and burdens.

These information streams will be exacerbated as people add sensors not only to their external possessions but also to their bodies. It is easy to imagine GPS devices in our children's clothing (no more lost kids in the shopping mall) and it is a small step to ingesting or implanting monitoring devices for our blood pressure, cholesterol, insulin, and other physiological processes. The streams of data such sensors acquire must be processed and possibly stored. Whether we add body LANs as another layer of network or not is not so important as how we configure, control and manage these new sources of data. A brief reflection on how targeted advertising has evolved based on streams of data from grocery store purchases is suggestive of the potential value of knowledge about our personal needs and habits. Weiser's vision of ubiquitous computing evolved to a vision of calm technology that is transparent to users. Although I see the technology itself fading into the background, I am less sanguine that the information management problems will be anything resembling calm. Although using telephones to make long distance calls has become trivial, choosing a long distance carrier or understanding one's phone bill(s) has become more complex.

If bit streams reflecting ever more detailed slices of our lives are to be secured, leveraged, and safely disposed, we have essentially a library problem. The need for information specialists who develop and help us use the conceptual tools for our personal information databases and libraries is clear. Additionally, many people will choose to employ information specialists to directly configure (and maintain) and possibly manage their personal and public information spaces. Although they may not be called librarians, the information specialists of the new century will trade as much upon their integrity as their understanding of human information needs and technical skills. Moreover, information specialists will find themselves between the technical and commercial forces that aim to maximise use of personal and public information and those who react to this trend. Issues of privacy and information overload will surely lead some to disconnect from the global information infrastructure. The wise information specialist will not point out what is missing in these lives but rather learn from them what qualities and strategies are most beneficial to foster (intermediate) human to human dialogues so that understanding and mutual respect prevail. This is a new mediator role for information specialists – balancing the information needs of those fully immersed in the global information infrastructure with those who choose not to be and balancing the need for secure private information with the global hunger for public information.

Personalisation

A second trend is toward more customized and personalized information services and systems. This trend is a result of larger portions of the world population having access to IT. The term 'universal access' is used to describe a world where everyone, regardless of language, culture, or physical capability can opt to access the global information infrastructure. There is an inherent tension between the standards that drive communication, physical exchange, and system interoperation and the individual needs and preferences of billions of people around the globe. It appears that standards are becoming part of the global information infrastructure while local clients are supporting increasing levels of customisation. Although some see these local preferences instantiated with personal software agents, I believe that alternative interfaces defined and controlled by individual users will play crucial roles in personalizing our electronic work and play environments. Current applications and browsers provide various kinds of preference settings and as research continues to understand how user characteristics and information needs are identified and mapped to conceptual interfaces, designers will provide richer arrays of alternatives to allow people to best meet their information seeking needs and to express their individuality electronically.

It seems clear that information specialists will again be called upon to play personal configuration management roles in personalized IT systems. However, information specialists have an even more important intermediation role in the larger global information infrastructure as the trend toward personalisation is balanced with the complementary trend toward special interest groups and distributed communities. Just as humans have a need to differentiate themselves from others, we also have a basic need for community. This is a more basic human dichotomy than the public/private dichotomy resulting from ubiquitous access and is best reflected by the impact of Internet news groups, chat services, and the open source software movements. People want to be distinguished (and rewarded) as creative thinkers and actors, while contributing to the welfare of their friends and community – open source contributions (including advice on newsgroups or chat rooms) accomplish both these goals by serving both ego and altruism needs simultaneously. Global networks foster this by allowing the critical mass of special interests to accrue quickly beyond the constraints of time and space. Information specialists have important roles to play in helping such communities to form and in managing information flow and exchange with the larger population beyond the community. This includes helping clients find not only the fruits of online communities but carrying the distinctions and culture along so that clients can smoothly join and leverage the community.

Microcharges

A third trend is less well developed at this time but seems inevitable as e-commerce and WWW access continues to evolve. Microcharging has yet to take its place as an economic model for information management. The current models are simple purchase (e.g. buy object/unit via credit card), public funding, subscription, and advertising, with the advertising model driving current WWW commerce for information services. At present, the costs of managing small units of information and tiny transactions exceed the individual value of those units and transactions. This will surely change as the potential profit of even miniscule (e.g. ten-thousandths of a dollar) charges for the trillions of online actions each day drive a tracking infrastructure that gives tighter control over smaller units of information. A significant portion of the voice telephone circuitry is already devoted to payment information and although the early proposals for electronic cash were not immediately adopted, microcharges for Internet activity are inevitable. Highway systems are common metaphors for electronic networks and the physical highway system also serves as a tangible example of the trend toward practical microcharges. Toll-booths are strictly limited because they interrupt traffic flow and require personnel and capital costs to maintain. Smart tags that attach to windshields are commonly used on busy toll roads and it is conceivable that such tags will be built into vehicles (consider unique identifiers in each microprocessor). Together with sensors built into highways, such technology will enable per kilometer charges for travel in larger portions of the overall system and could eventually change the highway system from mainly a public infrastructure to a hybrid of public and usage-based system. The political and social debate over such developments will surely involve technical, economic, and political leaders and should include information specialists as well. More specific to this chapter, microcharges on Internet activity will create a host of information management challenges that will require well-prepared information professionals. What one pays for (what is free and what is fee), how one pays (by the page for web sites or by bit rate) are only some of the issues that go beyond the capture, storage, and processing of microcharge information. A consumer-oriented perspective is a natural position for information specialists who will balance the values of free (publicly funded, open source), ad-driven, and fee-based information. Thus, information specialists must go beyond the technical issues of storage, routing, and retrieval to focus on the social and political balances across political, cultural, and economic boundaries.

The reactions to microcharges will include obvious solutions such as wholesale/bulk-rate services, but the complexity will surely lead to emergent conditions that will demand creative and human-centered responses. We can imagine microcharges pushed to capture machine cycles and many variants on manifestations of human cognitive cycles (e.g. mouse clicks, eye gaze). The alternative, advertising-driven services with targeted marketing is the current trend, but in the hybrid world of tomorrow, information specialists will need to balance the pros and cons of these competing economic models for information access to best suit the needs of clients and society at large.

Conclusion: Emergence and Finding the Balance

The trends outlined above involve multiple, complex systems and will surely lead to emergent conditions, objects, and behaviors. Although we can speculate about what these new phenomena will be, we cannot optimally plan solutions. Information specialists must build upon first principles related to human information needs and behaviors to balance the powerful forces that technology exerts on our lives. To do so, information specialists must first study and understand human information needs, information-seeking strategies, and the nature and structure of information as the abstractions of existence. Information specialists must, of course, also acquire the knowledge and skills to understand and apply IT to information problems. Because IT advances so relentlessly, this is a lifelong learning challenge. Third, information specialists must develop tolerance and objectivity for the wide ranges of diversity the global information infrastructure implies. Tolerance is as much a belief as much as a rational effect of understanding. It is essential because it is the basis for the trust and confidence that lies at the heart of the information specialist's professional value. This value is traditionally expressed in selection and quality control, information packaging, and equitable access – each of which will become more important in the emerging world of ubiquitous access, hyperpersonalized information services, and mixed economic models. It is what keeps our fulcrum sharp.

The information specialist is an intermediary, a counselor, a bridge, and a fulcrum – balancing the competing needs and forces of people and technologies. As such, information specialists balance information needs and information overload; personal preference and IT pressure, human self and genetic pool; and knowledge and belief. There can be no more exciting and challenging vocation in the new century.

21

Experiential Documents and the Technologies of Remembrance

Clifford Lynch

There is a rich selection of traditional literary genres that subjectively record events and experiences: histories and travelogues, biographies and autobiographies. There are also well-established objective forms. Perhaps the most commonplace is the transcript of legal or legislative activities; more specialised is the move-by-move transcription of a game such as a chess match. The visual arts, notably – drawing, painting, and sculpture – were employed to complement literature by more or less objectively depicting people and places within a representational tradition as part of the capture of experience. This has been the basic underpinnings of western civilisation's cultural record since the Renaissance, though the roots of these practices are much older.

These materials fill our libraries, museums, and archives, and form much of the record of our social discourse and cultural history. The technologies of publishing have made literary works widely available. Libraries both preserved them and helped them to reach the public. Some products of the visual arts were also mass-produced and distributed. Creators working within these traditions – and the publishers who served them – generally welcomed the idea that their works would endure.

There was a vast body of activity that could only be experienced as performance of one sort or another and subsequently remembered; technologies to capture performance did not exist. This body of activity, captured by the term 'performing arts', included theater and dance, as well as oratory, historical events, and indeed a very broad range of human experiences. There were notational schemes – scripts or musical scores – that permitted certain specific genres of performances to be *reproduced*, or *restaged*, but only with great loss of detail of the specific experience. These notational schemes were open to tremendous interpretation by the directors, conductors, and performing artists. And observers, recording these performances in words, only offered a limited, subjective, second-hand sense of what it was like to actually experience them.

The 19th and particularly the 20th centuries have given us a wide array of technologies that allow us to capture and record events and experiences, relatively objectively, and with great detail and fidelity: photography, sound recording, motion pictures, and video (haptic and olfactory experiences remain beyond the easy reach of current technology). One can argue that these technologies have enabled profound changes in the arts: by removing the demands to represent actual events, painting has moved into a variety of abstract schools; without the need for the musical score as a representation of a work, improvisational musical genres have become commonplace. Recording technologies have become very sophisticated and pervasive to the point where routine capture of sound and images are simply a part of our day-to-day environment and any consumer can casually employ them. Indeed, we deploy cameras as sensors – surveillance devices – that record constantly just in case an interesting event may take place in front of them.

Recordings made using these technologies speak directly to our senses and are very powerful surrogates for the experiences that they capture. Libraries, museums, and archives have begun to collect these materials, though less extensively than the printed word. Despite their power, they are often viewed as somehow less important than traditional literature or paintings; as new genres they are marginalised and distrusted. In addition, as our technological capabilities to record have improved, these works have presented extraordinarily complex legal and social problems as society has tried to sort out the various intellectual property rights of performers, authors, recorders, and the public at large. As content becomes divorced from the artifacts that temporarily store it, legal limitations have increasingly constrained its use. Finally, the economic models used to distribute these materials have moved away from the simple sale of artefacts like books that could be inexpensively acquired and placed in personal or library collections. Consider, for example, the distribution of films to theatres before videocassette recording, or works that are broadcast on radio or television. In many cases they simply weren't available for acquisition.

The cultures of broadcast radio and television, of Hollywood, and to a lesser extent of music marketing have moved far from the symbiotic relationship and sense of common purpose that has developed among authors, publishers, libraries, and readers. These new cultures do not much honor preservation of artifact within the cultural record as a public good.

This transition has led to a strange situation. In the past, when there was no way to record a play or a musical performance, the public had no expectation of anything but memories and second-hand descriptions. Today, it is possible to broadcast material (such as the nightly news) or to distribute a major film to theatres which reaches millions of people and consequently is of great social significance. But while recordings exist, they have no natural way to enter the public sphere managed by libraries and archives which ensure preservation and ongoing access by the public. Pay-per-view business models for digital information are extending these trends. We once understood performances as experiential works because we could not capture them and convert them to artefacts; today experiential works are created by marketing strategy rather than as the result of limited technological capabilities.

In the United States, copyright is a bargain between creators and the public at large. Creators are given incentive to create by the assignment of rights that they can exploit economically for a limited time. In exchange for these rights, the material ultimately enters the public domain. Rights are subject to fair use and other limitations, and historically there has been a presumption that publication would permit the work to become part of the broad cultural record and be preserved, either through copyright deposit or the actions of libraries in purchasing and archiving the work. This bargain is unraveling rapidly with the extension of copyright terms to unconscionable duration, and the increased distribution of works as experiential goods. The scope of intellectual property is also expanding, from actual creations to characters, imagined worlds, styles, and personas.

We are also becoming increasingly skilled at generating images, sounds, and moving images synthetically rather than using these technologies to record events. We use recording technologies as media for new art forms and vehicles for communicating and conveying information. The results of these growing skills are both exciting and sometimes profoundly disturbing as we still too frequently assume that sounds and images are recorded rather than synthesized. Hybrid forms that combine recordings of actual events and people with computer-generated characters and events are increasingly commonplace, and are calling into question the roles of images and sound recordings as evidence, record, and property within our society. And these new forms of art and communication, because they are carried within technologies of recording rather than writing, are again being offered as experiential works rather than artifacts or abstract content that can be sold for repeated viewing or loan.

As we stand on the threshold of a new millennium, technology has now given us the means to capture many types of events, experiences, and artistic activities permanently. But technology is also producing a whole new array of materials that may be initially presented as experiential works but that are simultaneously digital artifacts that can be stored, archived, annotated, and subsequently shared. These materials will raise profound legal and intellectual problems. Interactive conversations of any kind conducted over computer communications networks – interactions as simple as chat sessions or audio or video conferences or as sophisticated as a group of scientists coming together in a 'collaboratory' space to capture data from scientific instrumentation, review simulations, and perform analysis – are potentially both temporal activities *and* permanent works in the digital environment. As we interact with simulations, navigate through three-dimensional virtual reality scenes, or play computer-based games, we have the ability to capture not only the system itself for future interaction but the actual interactions with the system. It is not dissimilar to recording chess games as well as the rules of chess and a chessboard, though the level of complexity is enormously higher.

Before the computer-communications environment became pervasive, recording of events or activities generally required a conscious, deliberate commitment, an action. (Such actions have progressively required less and less effort and cost. It is now routine to videotape every meeting of certain government bodies or every individual passing through certain checkpoints, or to record every performance of a musical group.) In the digital environment, the best we can do is to beg the computer systems involved not to make a permanent record of an interaction that they mediate – and we are learning to have little confidence that these requests will be fully honored. And, if history is any guide, we will develop synthetic art forms that can also be delivered in these new artifacts, as well as using them to record activities, experiences, and events.

The retention, reuse, management, and control of this new cornucopia of recorded experience and synthesized content in the digital environment will, I expect, become a matter of great controversy. This will include, but not be limited to, privacy, accountability and intellectual property rights in their broadest senses. And these materials will hopefully become an essential and growing part of our library and archival collections in the 21st century – particularly as we sort through these controversies. They are very different from printed works, and different from the older recordings of events in the physical world. They are sprawling and complex, and we know even less about how to index and organize them (and how to apply technology to these tasks), and about what is important to keep and what is not than we do about images, sound, and video recordings of events. They stand completely outside of our existing systems of publishing or even our existing business models for the products of recording technologies. The economic and ownership frameworks for these new genres have yet to be established. Because of their intense and intimate interdependence with the computer systems that created them, their long-term archiving is also tremendously problematic.

The computer-communications revolution is sweeping up our technologies of remembrance and recording into a vast digital convergence. And these are only the superficial effects of the revolution. Certainly they will be disruptive to our cultural institutions such as libraries and archives, to our ideas about intellectual property and records, and to the economics of creation and publishing or dissemination in many fields. But at a millennial transition it is appropriate to look further ahead, to speculate about new technologies of remembrance that may emerge and to antici- pate their implications. While today these possibilities are the province mainly of science fiction writers (and they have been compellingly, thoughtfully, and beautifully explored in that literature), they are per- haps no more fantastic than the concept of a portable video camcorder and digital video editing with computer-generated special effects would have been to the average person in 1800.

In capturing and reproducing sound and images we mimic the human perceptual apparatus of sight and hearing. We are at the very beginning of developing assistive technologies for the deaf and blind which replace the *output* rather than the input of these perceptual systems. Rather than generating sound or light, we generate the output of the systems that sense sound or light. As we learn to capture and regenerate the output of human perceptual systems – not just sight and sound but touch and taste and smell – as well as their input, this will allow us to record a far wider range of human perceptual experience, and to render it for subsequent consumption in a very compelling fashion. As we gain the ability to syn- thesize our senses, much as we synthesize sound and images rather than simply record them, we will enable entirely new forms of communication and art as well as the recording of experience. At another level, however, I would argue there's nothing fundamentally new here, except perhaps for the introduction (or re-introduction) of greater individual subjectivity into the technologies of recording: perhaps we will see people make a living as particularly skilled and sensitive viewers or listeners or experiencers. But we will still capture, archive, and share these record- ings. They will be part of library collections, along with our sound and video recordings and our written words.

In the next centuries we will perhaps learn to extract, store, and share human memories. The distinction between capturing perceptions and memories is complex and philosophical, and perhaps has to do with when the recording is made and what is being recorded – whether it is contemporaneous with the perception or obtained after the fact – and the extent to which the perceptions are subjectively reshaped. And while it's at least somewhat plausible to restrict (real-time) capture of perception of an event someone attends, much like events restrict other recording de- vices today, it's very hard to believe that we will be able to impose lifelong restrictions on the sharing of memories of all the experiences an indi- vidual has enjoyed. Can we even think about establishing rights *to be remembered*? It is not difficult to envision recorded memories becoming part of the cultural record and being managed by cultural heritage insti- tutions. It is also not difficult (although it is disturbing) to imagine a commerce in memories – though it is not clear whether it will be one of experience or of permanent copies.

We are beginning to learn to simulate the human creative process, at least in limited domains, by computer; we are learning how to capture behavior. As we develop systems that can write concertos in the style of J. S. Bach, improvise in the style of Miles Davis or Jerry Garcia (within a context of other participating musicians), or write mysteries in the style of Agatha Christie, we will find that our social understanding of creative works – perhaps most prominently improvisational creative works – may become much more complex. Recognise, of course, that such simulated creative processes aren't limited to the arts; the broader issue is one of ever more capable, accurate, and detailed computer simulations of various aspects of the capabilities and personalities of individual people. Will computer-generated personae become cultural artifacts? Will libraries collect them? How will we choose to balance the authorship rights of those who build these simulated personas and the rights of those being simulated?

And beyond that, at the speculative horizon, is the possible ability to capture snapshots of actual consciousness, which moves us into the realm of science fiction. Such a technology, if it proves to be possible, raises the most basic issues of identity and immortality. Here we are dealing not just with a computer simulation of some of a person's capabilities or behaviors, but in fact a way of capturing the essence of a person at a given point in time. This is in some sense the ultimate technology of remembrance, where the acts of living and thinking take on an artifactual persistence. The social and cultural implications of such a technology are so profound and wide-ranging that it is hard to think about any of our current cultural heritage institutions as doing anything except maintaining the records of the time before the technology, or of communications artifacts that may continue to be produced.

All of these future technologies of remembrance promise greater capabilities to mechanically translate experience and behavior, and perhaps even living and thinking, into artifacts. Recounting experience or memory is no longer an artistic gift in this possible future. Such developments point towards a massive breakdown of our current ideas about intellectual property, to a time where the experience or even the memory of the experience of an event captures that event, and in which creation of a purely experiential work implies ongoing control over those that experience it. The experiential works that have become so problematic in the 20th century as we have begun to develop the means to capture them into artifact will become irrevocably artifactual in the next millennium as the technologies of remembrance are perfected.

There are extraordinary possibilities for new art forms and genres of communication enabled by the technologies on the horizon, as well as opportunities to capture and share a much broader part of the human experience. The new technologies of remembrance will raise fundamental questions about what it means to own and disseminate intellectual and creative works, to maintain a record of cultural heritage and social discourse, and how we will structure and assign missions to the institutions that maintain these records.

Acknowledgements and Sources

I was asked to prepare a brief, speculative vision of how information may evolve in the future for this collection. While detailed footnotes do not seem appropriate, I need to recognise that the ideas presented here draw from, and sythesize a very wide variety of sources. In particular, I need to acknowledge the influences of William Mitchell's book *The Reconfigured Eye: Visual Truth in the Post-Photographic Era* on the effects and implications of photography and of Hans Moravec's *Mind Children* and Ray Kurtzweil's *Age of Spiritual Machines* on the possibilities and implications of the new technologies of remembrance. Science fiction writers have explored these areas long before the scientists arrived. While a complete list would be impossible, key sources include John Barnes' *Mother of Storms* on capturing perception, the works of Cordwainer Smith and Daniel Keys Moran, Greg Bear's *Eon* and *Eternity* on downloading memories, and George Alec Effinger and Phillip K. Dick (particularly Dick's story 'We Can Remember It For You Wholesale'). As I've struggled with some of the ideas expressed here, conversations with Brewster Kahle, Wendy Lougee, and Cecilia Preston all helped in clarifying aspects of the issues. Finally, Nancy Gusack did a tremendous job of editing and sharpening the ideas here.

22

Waking the Giant – the Internet and Information Revolution in Africa

Mike Chivanga

Introduction

Information and Communication Technologies (ICTs) – in particular the Internet – have been identified as major keys to Africa's socio-economic development (Mandela, 1998). These expectations are being echoed at nearly every conference, seminar and exhibition under the over-popularised flagship of change flamboyantly coined 'Africa's Renaissance'. Africa's hopes of a socio-economic development that has been tried and tested by the use of a horde of socio-economic models is once again being wooed to these new technologies as keys that can consequently lead to change. While one can identify many ways that the new technologies can spear-head the resuscitation of Africa's fortunes, there are many hurdles that still have to be overcome. The Internet, for example, has the potential to reduce the information gap between and within Africa, provide access to cheap and fast communication crucial not just to economic development, but also to the speeding up of the democratisation process that is gradually gathering momentum. Information and knowledge delivered via the Internet are undeniably linked to socio-economic progress (Wolfensohn, 1998). Many African countries are still to leap into the Information age and Internet technologies present this opportunity more than any other technological force before it (Peter Knight, 1995).

Suggestions will be made as to how Africa can reap the benefits that are being offered by new technologies like the Internet. A proper seedbed has to be cultivated that will ripen the fruit being offered by the Internet and accompanying technologies. New technologies have the greatest potential to develop information societies – the cornerstone of economic success. Firstly, the key to the information revolution in Africa lies in having the right telecommunication infrastructures. Tied to the development of the infrastructures are other factors that have to be considered as shall be discussed in the following sections. The second key is developing information resources that can have an impact on all socio-economic activities.

Reference will be made to developments in China where a majority of the people are now using new technologies in their everyday lives and see their significance in the process of social and economic change.

The Internet and Development

Internet technologies have the potential to disseminate information that can impact on development. The web, for example, makes it possible for individuals and organisations to have equal access to information more than any other technology before it. Many people in Africa are shut out of the global communication belt because they don't have the means to make use of it. This is seen at a basic level through media censorship. At an infrastructural level, a majority of the people (about 80%) don't have access to telephones and electricity to make the necessary Internet connections (Muellar, 1999).

What one should note is that a majority of the people have done nothing to deserve the state they are in, i.e not to have access to information that they can independently use for decision-making. From colonial times through the period of independence, the political and economic systems have ensured an allocation of resources to serve the elite and the upcoming middle class. What one would have hoped for after political independence was honest leadership committed through deliberate actions to speeding up the development of infrastructures essential in leap-frogging many societies into the Information Age. Development priorities include better housing in rural areas and schools that have telephone lines and electricity necessary for the installation of computers that are ready to be plugged into the Global Information Infrastructures (GIIs).

Internet Developments in Africa

The estimated number of Internet users in Africa up until March 1999 was about 1.2 million. It is very difficult to measure the actual number of Internet users. Realistic figures were gleaned from ISPs in Africa based on the number of dial-up accounts registered with them. The tally came to slightly over 400,000 subscribers in the whole of Africa (Mike Jensen, 1999). According to a recent study undertaken by the UN Economic Commission quoted by Mike Jensen in his yearly status report, each computer which supports Internet services like accessing the web and email has an average of three users. This gives us an estimated number of Internet users in the whole of Africa of about 1.2 million. 'Of these there are about 700-800,000 in South Africa leaving 400-500,000 amongst the 734 million people in the continent'.

Many countries with low Internet users have serious political and economic problems, some of which will take over ten years to completely resolve. Examples include, among others, Burundi, Liberia, Rwanda, Congo, Somalia and Eritrea. Three-quarters of the countries have poor telecommunications infrastructures essential for the growth of the Internet. It will again possibly take between ten to twenty years for the telecommunication systems to improve. Fibre-optic cabling is underway and countries like Bostwana, Mozambique and Zimbabwe are already benefiting from this infrastructure that facilitates high-speed transmission of Internet data. The model being followed by many African countries however should also make the development of data communication infrastructures a pressing concern as they have better capabilities to support Internet traffic than voice communication systems.

Lessons for Future Developments

China presents a challenging example of how ordinary people are becoming more technologically savvy at an unprecedented rate in the country's history through active participation by the country's leaders. Away from mainstream politics, China's political leaders 'consider the *PC an essential modernization tool*' (R.Wallace, 1999).

The Chinese model is worth emulating from an African point of view as a majority of the political leaders, unlike in China, are standing in the way of progress. Institutional corruption, technological ignorance and the fear that freedom of information can jeopardise their political futures are just some of the hurdles that need to be overcome for the true Net effect to happen.

The rate of change according to industry analysts is that 'China is connecting more than 20 million new telephone lines every year, installing fibre-optic trunk lines and cellular telephone systems as well as building a domestic satellite network. In China, even the taxi drivers know about the Internet' (Mo Krochmal, 1999). Even though the Internet market is still young, the building of the necessary infrastructure continues unabated.

The first tenet of the model that can used to increase Internet connectivity (eventually leading to the flourishing of information services that can impact on development initiatives) is a shift in the attitude, behaviour and operations of political institutions. The direct involvement of political leaders in the process of technological change in China has resulted in a tacit acceptance of their revolutionary potential. While political institutions in that huge country are facilitating the proliferation of the Internet services essential in fanning an information-aware environment, many African countries are responding very slowly to this new phenomena. Political chaos leads to poor economic performances in any country and this has a direct bearing on investment and growth in many sectors. In Africa, this trend is not likely to change for many decades in the next millennium. Governments that have been in power for many years and not wanting to relinquish power is a typical problem. Ethnic atrocities leading to unending feuds over borders, and corruption on a transparent scale are just some of the problems that have to be dealt with.

There is a direct correlation between countries with low numbers of Internet users and unstable and oppressive political regimes. Governments should actually see the Internet as a tool that can be used to inform and educate the public on many issues. Participation in political decision-making by a majority of the people will eventually lead to stability – highly significant for economic growth.

Milton Mueller (1999), the director of the graduate program in telecommunications and network management had this to say about developing countries (equally applicable to Africa): 'If developing countries can remove the political and economic barriers to the growth of the Internet, the economic opportunities for growth will be huge…. If you don't have roads or electrical power, it's hard to do anything with a computer.'

The second tenet of the model essential to the development of information societies in Africa, as pointed out earlier, is having the necessary infrastructures. In many developing countries 'voice communications is still the main driver of communications technology' (Mugo Kibati, 1999). Development of data communications should be a major concern as much as voice communications. This should be seen in the regulation of the telecommunications industry. Unfortunately there is an unhealthy monopoly in the industry with governments controlling the regulation and supply of telecommunication services. These state controlled telecommunication organisations have contributed to the uneven distribution of telephone lines between rural and urban areas. A privatisation of part or the whole of state-owned PTTs is crucial in the development of a competitive market that can spur the development of the Internet.

Another aspect of the information revolution in Africa is the development of content that is in tune with socio-economic conditions. A large majority of the information resources found on the web originates from outside the continent. This bias will certainly not address the information gaps found in the continent. Individuals and organisations should be educated about the value of information, its organisation and management. Indigenous people have so much information about their environment that is not properly archived. The development of content should start with what they know and fuse it with information generated from overseas thus laying a solid foundation for the building of information and knowledge societies.

In conclusion, there are three factors pivotal to the future development of Internet technologies crucial to the nurturing of information societies in Africa. Firstly, there is an urgent concern to change the political systems. This will pave the way for the development of a regulatory environment that can introduce, develop and maintain change that comes through technological influences like the Internet. Economic success will also follow, consequently resulting in a majority of the people having the financial means to afford the new information services. Secondly, the development of the necessary telecommunications infrastructure will naturally lead to an exponential growth of the affordable Internet services. Last but not least is the need to produce information resources that takes into account the unique socio-economic characteristics of communities in the continent. There has to be an active willingness to change. At Africa Telecom 98, the Secretary General of ITU, Dr Pekka Tarjanne, very correctly pointed out that one of the biggest obstacles to change are people. He put it succinctly, saying, 'The human animal does not embrace change easily or willingly. We are by nature conservative creatures, who often prefer to live with familiar methods and frameworks, even if they are difficult or disadvantageous, rather than take the leap into the unknown.' (Dr Pekka Tarjanne, 1998).

References

Mike Jensen (1999). 'Africa Internet Status', http://www3.sn.apc.org/africa/afstat.htm

Peter Knight (1995). 'The Telematics Revolution in Africa and the World Bank Group', http:www.knight-moore.com/html/telematics_in-africa.html

Mugo Kibati (1999). 'The Wireless Local Loop in Developing Regions', *Communications of the ACM* 42 (6), June 1999.

Mo Krochmal, 1999. http://www.techweb.com/news/story/TWB19980514S0017).

Nelson Mandela (1998). Opening Speech, Africa Telecom 98, http://gold.itu.int/telecom/aft98/index.html

M. Muellar (1999). http://www.techweb.com/wire/story/twb19990609S0026

Dr Pekka Tarjanne, (1998). http://gold.itu.int/telecom/aft98/index.html

R.Wallace, 1999. http://www.techweb.com/wire/finance/story/INV19980426S0001

James D. Wolfensohn (1998). Knowledge for Development, World Development Report

23

Toward a Sustainable Science of Information

Amanda Spink

A Science of Information

Information is usually associated with something else. Information doesn't exist by itself. Information may be an attribute of a text, image or other artifact. Information may be a cognitive human construction during human information behavior processes. However, the study of information is not the study of information itself, but the study of a human effect called information. We can only really see information in terms of the behaviors of humans as they generate and sustain a process of constructing information during seeking, retrieving and use process. A science of information is concerned with studying such a phenomenon. Our thinking about a science of information is always evolving as we approach thinking about information at various levels and models have developed at different levels. Our thinking about information is evolving within a cognitive view, an organisational view and a societal view. How these levels mesh or dwell together is not clearly understood.

To begin, I examine: *What can be the nature of a science of information?* A science of information involves the study of complex processes and many human information-related activities, such human information seeking and retrieving behaviors, and associated organisation of collections of texts, images, sounds or multimedia; the intellectual representation of such texts derived by humans directly or indirectly by algorithms; and intellectual ways and means of searching and retrieval by users; and facilitating systems and techniques. The complexity of human information seeking and retrieving is derived not only from these complex processes, but from the direct involvement of human in the generation and user of texts in information systems, with associated cognitive, affective, social and situational (problem, task) variables. In other words, a science of information is not only a technical but a cognitive, social, and situational process. With the marriage of computers and telecommunications, a science of information is concerned with facilitating human information behaviors.

A science of information is also presented in a problem-solving context as humans' experience information problems on an individual, organisational and societal level. Different processes have been designed to allow humans to construct information at these levels. Libraries and the web are good examples of systems and processes set up to help humans to construct information as individuals, organisations and societies as a whole. However, humans experience information overloaded from the systems and processes they have created with associated trouble organising, retrieving and using information. The web has also made life one big information headache! Information is also now so linked to technology. Increasingly we don't separate information from technology. However, information as a human phenomenon can be isolated from technology, as information is basically a human construction not purely reliant on technology.

As scientists of information struggle with these issues, a further challenge is emerging that necessitates the linking and exploration of the relationship between two previously unrelated concepts – *human information behavior* and *sustainability*. Many elements are increasingly binding the concepts of information and sustainability. A science of information needs to explore the fundamental relationship between human information behaviors and sustainability at the cognitive/behavioral, organisational and societal level. These processes may provide some building blocks for a general framework for a science of information. Scientific fields are driven by the need to solve certain problems and explore certain philosophical issues. A science of information is driven by the need for human and technological solutions to the human information overload problem. The emerging problem of human information behaviors and sustainability extends the approach that views the science of information as concerned with problem solving.

Sustain as a concept means to keep from falling or sinking or failing or enabling to last out. A science of information focuses not only on the technology and structures of information collection and delivery, but also on the nature of human information behaviors and how they can be kept from failing, sinking, falling and enabled to last out. Humans often have difficulties sustaining effective information behaviors at cognitive, organisational and societal levels. An implication of this difficulty is that many human information behavior processes supported by systems and processes are not effective. Humans need enabling systems to allow them to persist in information behaviors that lead to more effective outcomes. The process of solving various levels of information problems involves developing ways to sustain human individual, organisational and societal information behaviors and problem-solving processes. A science of information needs to investigate the relationship between sustainability and effective information behaviors, and the development of *sustainable* information-related process at all three levels.

The first level of investigation for such an endeavor is the area of human information behaviors at the cognitive/individual level.

Cognitive/Individual Level

At the human cognitive theory and behavior level, we need to help individuals solve their information problems and sustain their human information behaviors. Sustaining human information behavior processes such as information seeking processes may include interactions with IR technologies. From childhood, humans develop the ability to seek and use information. Within these processes they develop interaction skills to retrieve information from IR systems such as online public access catalogs, the web and digital libraries. However, research shows that many human interactions with IR systems, such as digital libraries and the web are short and ineffective (Jansen, Spink, Jansen and Saracevic, 1998). Humans often do not conduct persistent information seeking and interacting, particularly when interacting with IR systems. We can ask the following questions: How do humans sustain their human information behavior processes? Why do humans often not persist in many instances when seeking information? How can human information seeking, retrieving and use processes be sustained? Why must they be sustained? Why is the sustaining processes important?

The sustaining aspects are important because human information behaviors, such as information seeking and interactions with IR systems, are complex and cognitively demanding processes that must take place over time to be effective. We know that many people browse as an information behavior, preferring to wander and serendipitously meld through information spaces. Browsing can be sustained for longer periods of time than more deliberate searching. However, many tasks require more deliberate searching and more complex interactions with humans and processes or systems. Sustainability includes depth as well as breadth and time. To understand your information problem and articulate and pursue it in sufficient depth and breadth in a manageable timeframe is a key element in human information behavior. Of course, this is based on the assumption that more effective information skills are an increasing crucial element of a successful role in the information or knowledge based economy.

From the model we can see that a science of information focuses on many different processes that occur over time and need to be sustained. These processes include: a human information problem that initiates human information behaviors related to a human problem state, cognitive state and knowledge state (Ingwersen, 1992, 1996); an information seeking behavior (Kuhlthau, 1993); they may include human interaction with IR systems (Saracevic, 1996,1997 and issues such as feedback (Spink, 1998b; Spink and Saracevic, 1998) and relevance (Spink, Greisdorf and Bateman, 1998). In particular, the concept of *human information coordinating behavior* (HICB) is an important linking and sustaining process for a science of information that binds together the many processes involved in human information seeking and retrieving. The next section of this essay proposes a discussion of the key sustaining process of HICB.

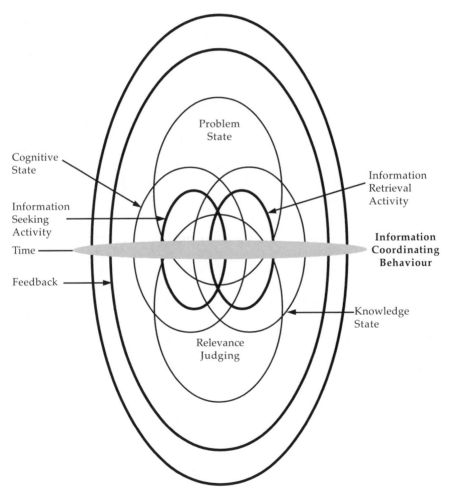

Figure 1 presents Spink and Greisdorf's (1999) model that depicts many human processes of concern to a science of information.

Human Information Coordinating Behavior

Humans' *coordinate* a number of elements, including their cognitive state, level of domain knowledge, and their understanding of their information problem, into a coherent series of activities that may include seeking, searching, interactive browsing and retrieving and constructing information. The development of a science of information necessitates a theoretical and empirical explication of the nature and role of HICB.

There are two levels of HICBs –

1.   Information seeking level
2.   Information searching (HCI) level

Humans cognitively coordinate their information seeking level behaviors with their information searching level (human-system interaction level) behaviors, including the recognition and making sense of and cognitively articulating an information need or a gap in their knowledge. Humans then coordinate these processes to construct an information-seeking process. Coordination is also related to movement through a human information-seeking process. Humans' information coordinate in order to move through their information-seeking process. Part of the information-seeking process is the translation of their information problem into a form that allows them to construct information from texts in the broadest sense. Bringing the elements of their information problem to an effective information-seeking and searching process is essential to an effective co-ordination process.

For example, a human is seeking information on a particular disease. They enter a library or begin to search the web. To enable their information-seeking and searching process to move forward they must understand the dimensions of their information problem and coordinate their information seeking and searching processes to the degree that they are able to interact with the functional structure of the library or web system. The coordination process between information problem and information-seeking/searching process must take place before a human enters a keyword into the web or begins to browse the library shelves. The output of the web search or the books found on the library shelves are coordinated through information feedback by the information-seeker with their information problem through various judgments of the relevance, magnitude and strategic aspects of the information system's output (Spink, 1997b).

Establishing and sustaining an effective information seeking and searching process requires humans to develop effective and coherent information coordinating behaviors and processes. In other words, an information seeker must coordinate a number of elements, including their cognitive state, level of knowledge, their understanding of their information problem, into a coherent series of sustained activities that may include seeking, searching, retrieving and using information. We know that hand-eye coordination is a physiological process that humans develop from childhood. But how do humans learn the process of coordinating their information needs into coherent processes of human information seeking and retrieving behaviors? An important element for the human information coordination process is information feedback.

The information feedback loop forms the basis of the looping HICB process. But, how do humans achieve coordination, including recognizing, making sense of and cognitively articulating an information need or gap in their knowledge, and construct and maintain an information seeking process? The informational feedback loop of coordination behavior is one of the most penetrating fundamentals in the information sciences and the basis for a theory of interaction. A basic issue confronting information science is the nature, manifestations and effects of feedback. The process of information feedback (Spink and Saracevic, 1998) facilitates human information seeking and searching (Spink, 1997a,b), and binds the human process of coordinating information-seeking and searching.

Effective matching has been seen as the key to effective interaction between human's and information systems. However, information systems must do more than match – they must assist users to coordinate, facilitate and sustain effective human information coordination behaviors. Information coordination is a fundamental human-initiated, facilitated and determined behavior. We are told that humans and information systems must collaborate. However, unless a human can coordinate, a collaboration with an information system may be ineffective. Human and information systems must also collaborate to facilitate an effective information coordination. However, coordination goes beyond collaboration – you may collaborate on one level, but not facilitate a coordination.

Human information coordination is a complex human process to model. Exploration of HICB is crucial to the development of a science of information and the design of more effective information systems. Information system designers are in reality trying to facilitate information-coordinating behavior that allows humans to effectively coordinate their actions within the context of their information seeking process. The effectiveness of human information coordination has been partly determined by relevance measures such as precision and recall, and can also be measured by changes in humans' stages in their information seeking process (Spink, Greisdorf and Bateman, 1998). New information system effectiveness measures may be based on human information coordinating behaviors.

HICB binds and draws together other processes at the heart of human information seeking and retrieving. A science of information can focus on many different processes drawn together by information coordinating behavior, including the relationship to other information-related processes. A growing number of researchers are already working towards a more integrated view of human information seeking and searching (Spink, 1998b). However, without the process of human coordination of the many elements of the human information seeking and searching process there would be no process. HICB is not currently an explicit process in models of concern to a science of information. However, its nature, manifestations and effects need to be explicitly investigated.

Information coordination is a rich notion of active human information behaviors. HICB lends value to Informatics research by bringing together interdisciplinary concepts and integrating them into a concise model and framework for ongoing studies and development. The explication of the process of human information coordinating behavior is fundamental to the development of a theoretical framework and model for a science of information. The modeling the human information coordinating process, in relation to human information behaviors of seeking and searching, may assist in the development of information systems that more effectively enable humans to effectively coordinate their actions within the context of their information seeking processes. We need to explicate the nature, manifestations and effects of HICB in the context of informatics processes.

This brings me to a brief discussion of the second area of a science of information at the organisational level in sustaining human information processes.

Organisational Level

At the organisational theory and behavior level, the role of information in sustaining organisations, organisational growth and productivity is an important area of research. Organisations adopt various information collection and dissemination structures and processes that range from libraries, to intranets and access to the internet and web. The goal for organisations is to develop effective sustainable information processes that enable organisational growth and development. However, most organisational models pay cursory attention to organisational information behaviors. Elements of environmental scanning and competitive intelligence gathering have been increasingly researched. However, the role of human information behaviors in sustaining reengineering and quality management processes is also in need of study (Prybutok and Spink, 1999).

The third area of concern to a science of information is the societal level.

Social Theory and Behavior Level

At the societal level, sustaining social information processes and the development of a sustainable information processes and societies is an important research area (Spink, 1995). In response to the problems of modernity, many social scientists and economists are projecting a future based on differing philosophical, social, political and economic perspectives on the nature of sustainable development. One view within the debate advocates sustainable development as a set of policies and methods to sustain global industrialisation, through sustainable technological and economic development.

An alternate view presents a scenario of global scarcities, conflicts and decreasing natural resources. The international economic and social transformation to modernity is increasingly associated with global population growth, migration to the cities, mass death, depletion of natural resources and energy fuels, and environmental degradation. In the future a downscaling of global industrial production to sustainable levels of development may occur based on regional self-sufficiency to preserve humanity. Sustainable development within this view represents a radical rethinking of the current social, economic and political dimensions of development. The scientific precursors to digital libraries, including computer and information science research, evolved during an expansionist period of unparalleled social, economic and technological development. This research has facilitated western industrialisation and information transfer within stages of an evolving information society.

Some information science researchers have begun to explore the dimensions of sustainable information services and technologies. Recent research funded by the International Development Research Centre (IDRC) of Canada focuses on the development and evaluation of information services and technology projects in LDDS, to strengthen the information infrastructure and industrialisation (Menou, 1993, 1994). The 47th General Assembly, Conference and Congress of the International Federation for Information and Documentation (FID) held in Tokyo, October 2-9, 1994 also called for more research in this area (Tocatlain, 1994). Sustainable information systems in LDCs (Agha and Akhtar, 1992; Johnson, 1991; Niang, 1993) and the role of agricultural libraries and library automation to sustaining LDC industrial development (Ashford and Hariyadi, 1993; Poitevein, 1992) have also been investigated. Further research is required to investigate the role and importance of information in sustainable development. The informational aspects of sustained global industrialisation and development, or the informational needs of an industrially downscaling society are also open for research. We are just beginning to explore the information needs of the alternate approaches to sustainable development.

Conclusion

This paper is but the beginning of the long journey of discovery. I have explored many areas and discussed many ideas related to human information behaviors at a cognitive / individual, organisational and societal level. This is a beginning framework for a science of information.

References

Agha, S. S. and Akhtar, S. 1992.,'The responsibility and the response: Sustaining information systems in developing countries.' *Journal of Information Science* 18, pp. 282-292.

Ashford, J. and Harriyadi, U. 1993. 'Library automation and sustainable information services in developing countries.' *Asian Libraries*, June, pp. 53-58.

Belkin, N. J., Cool, C., Stein, A., and Theil, S. 1995. 'Cases, scripts and information-seeking strategies: On the design of interactive information retrieval systems.' *Expert Systems With Applications* 9 (3), pp. 379-395.

Ingwersen, P. 1992. *Information Retrieval Interaction*. Taylor Graham, London.

Ingwersen, P. 1996. 'Cognitive perspectives of information retrieval interaction: Elements of a cognitive IR theory.' *Journal of Documentation* 52 (1), pp. 3-50.

Jansen, B. J., Spink, A., Bateman, J., and Saracevic, T. 1998. 'Real life information retrieval: a study of user queries on the Web.' *ACM SIGIR Forum* 32 (1), pp. 5-17.

Johnson, J. S. 1991. 'Computerizing information systems in developing countries: Keys to sustainable development.' *PAL* 22 (3), pp. 22-30.

Kuhlthau, C. 1993. *Seeking Meaning: A Process Approach to Library and Information Services*. Ablex Publishing, Norwood, NJ.

Menou, M. 1993. 'The impact of information on development: Results of a preliminary investigation.' Paper presented at the 3rd International Information Research Conference at Poigny-la-Forêt, France, 11-13 July, 1993.

Menou, M. (Ed.). 1994. *Measuring the Impact of Information on Development*. Ottawa, Canada: International Development Research Centre.

Niang, T. 1993. 'For a sustainable agricultural information system in French speaking Africa.' *Bulletin of the American Society for Information Science*, December / January, pp. 19-20.

Poitevein, G. 1992. 'AIBD: A valuable endeavor in agriculture in agricultural libraries and Information services for Latin America and the Caribbean.' *International Journal of Information and Library Research* 4(2), pp. 149-152.

Prybutok, V. and Spink, A. 1999. 'Using the Baldridge criteria for self-assessment assisted one health care provider with its quality advantage strategy.'*IEEE Transactions on Engineering Management.*

Saracevic, T. 1996. 'Interactive models in information retrieval (IR): A review and proposal.' *Proceedings of the 59th Annual Meeting of the American Society for Information Science 33*, pp. 3-9.

Saracevic, T. 1997. 'Extension and application of the stratified model of information retrieval interaction.' *Proceedings of the 60th Annual Meeting of the American Society for Information Science, 34*, pp. 313-327.

Spink, A. 1995. 'Information and sustainable development.' *Libri 45*, pp. 203-208.

Spink, A. 1997a. 'Information science: A third feedback framework. *Journal of the American Society for Information Science 48*(4), pp. 741-760.

Spink, A. 1997b. 'A study of interactive feedback during mediated information retrieval.' *Journal of the American Society for Information Science 48*(5), pp. 382-394.

Spink, A. 1998a. 'Toward a feedback model for informatics: Nature and manifestations of Information feedback.' *Proceedings of SCI?98: Systematics, Cybernetics & Informatics 98: Orlando, Florida. July 1998*, pp. 75-92.

Spink, A. 1998b. 'Toward a theoretical framework for information retrieval (IR) within an information seeking context.' *Proceedings of the 2nd International Information Seeking in Context Conference, August 12-15, 1998. Sheffield, UK: University of Sheffield, Department of Information Studies.*

Spink, A., and Greisdorf, H. 1999. 'Toward a theoretical framework and model for informatics.' *Proceedings of SCI99: Systematics, Cybernetics and Informatics Conference, Orlando, FL, July 31-August 4, 1999.*

Spink, A., Greisdorf, H. and Batemen, J. 1998. 'From highly relevant to not relevant: Examining different regions of relevance.' *Information Processing and Management 34*(5), pp. 599-622.

Spink, A. and Losee, R. M. 1996. 'Feedback in information retrieval.' *Annual Review of Information Science and Technology 31*, pp. 33-78.

Spink, A. and Saracevic, T. 1998. 'Human-computer interaction in information retrieval: Nature and manifestations of feedback.' *Interacting With Computers: The International Journal of Human-Computer Interaction.*

Tocatlain, J. 1994. 'A strategic alliance of international non-governmental organizations in information.' *Bulletin of the American Society for Information Science 21*(1), pp. 26-27.

Wilson, T. D. 1997. Information behavior: An interdisciplinary perspective. *Information Processing and Management 33*(4), pp. 551-572.

24

Networked Learning in Higher Education: An Agenda for Information Specialists

Philippa Levy

Introduction

Recent, far-reaching changes in the higher education environment make it a safe bet that technology will play a major part in shaping future educational practice in the sector. Already, the potential of new information and communication technologies (ICTs) to offer a variety of alternatives to physical co-location in learning activities and to reduce the need for trips to the library, is being exploited by programmes aimed at part-time, work-place and geographically distributed students. ICTs are also being incorporated into many traditional on-campus courses. The distinction between distance and campus-based learning is blurring and mixed-mode approaches – in which face-to-face interaction and traditional modes of information access are combined with the use of online communication and electronic resources – are increasingly the norm. National and international policy agendas promoting the principles of flexible, lifelong learning, coupled with economic pressures to deliver education more cost-effectively to larger numbers of people, point the way towards continued integration of technology into learning and teaching as rapidly as local circumstances will allow.

The task facing the sector as it moves academic practice further into the networked environment is considerable, and much depends on the capacity of institutions to resource and manage a complex process of cultural change – affecting underlying assumptions about professional purposes and roles, as well as organisational structures, work practice and relationships. As new models of learning and teaching are developed, much also depends on ensuring that good educational practice is put before technology. Within this context, the role for information specialists in networked learner support [1] – that is, in *multidisciplinary* support for effective networked learning and teaching, involving stakeholders in a variety of services and roles – is a critical area for strategic and professional development. Information services and staff have an important part to play in ensuring positive scenarios for higher education of the future. My aim here is to suggest that they need to become champions of *active* learning in the new environment, establishing an integral place for 'information literacy' within courses and at the same time contributing to a broader pedagogic agenda to develop good practice in networked learning.

A New Educational Space

It is already possible to identify all learners within higher education as 'networked', encompassing both the large majority whose use of ICTs is at present intermittent and ad hoc, alongside the small, but growing, minority who are enrolled on courses designed to take the fullest possible advantage of the network environment. A key feature of this is *convergence* between information and communication technologies: 'networked learning' refers both to provision of access to multimedia resources and to the interconnectedness of stakeholders in the educational arena, including students, tutors, information and IT specialists. There are already multiple modes of online interaction between people and numerous digital formats for the presentation of information content. Email, synchronous and asynchronous text-based conferencing, shared workspace tools and groupware packages, desktop and studio-based videoconferencing, can and are all being used to support learning. The web has rapidly come to be perceived not only as a means of information publishing and access but as an environment for learning, as evidenced by the availability of learning environment 'shells' which interface with, or are based on, the web.

Increasing emphasis is being put on the need to create more multimedia content for learning, and new software and standards aim to enable easy management of online learning resources, including resource-sharing across institutions and integrating content from multiple publishers. The wider electronic information landscape, too, is increasingly rich, offering a vast array of web pages of varying quality and many stable, high-quality electronic resources including full-text journals, library catalogues, bibliographic databases and numerous resources in graphic and video formats. Networked learners can tap into discussions on mailing lists, visit 'virtual communities', participate in exchanges hosted by a new generation of interactive electronic journals, or simply email subject experts directly in the hope of engaging them in debate.

No doubt another big chapter in the technology revolution will soon unfold. However, keeping up with what is already happening gives us at the grass roots level in higher education plenty to be getting on with for the foreseeable future.

It is clear that working within the new, 'converged' learning space – in the sense that it is at the same time a technology and an information environment, as well as an environment for independent and collaborative learning activity – students will have at their disposal an increasingly rich array of information and communication resources, many of them accessible via a single interface. To take full advantage of these for both formal and informal learning purposes, they will need opportunities to develop a range of new learning skills, especially in relation to on-line communication and group-work, technology access and manipulation, and the use of information resources and facilities. The emerging information landscape is challenging, and learners at all levels will need to develop the capacity to navigate it confidently, select resources that are appropriate to their purposes, and use them effectively in the context of their projects.

Active Networked Learning

The extent to which students engage with library resources and develop information skills depends largely on the influence of teaching staff and the way information issues are addressed (explicitly or implicitly) by the pedagogic models they adopt. At present, two distinct trends are apparent in the use of ICTs in teaching. On the one hand, models which use technology mainly to offer access to multimedia content embody 'transmission' philosophies of education, presenting online learning resources in packaged form and placing little emphasis on the 'process' dimensions of learning. This approach – often driven by economic pressures, and highly attractive to powerful new commercial players on the higher education scene – raises the spectre of a severely impoverished future for technology-based learning.[2] Add to this the strong tradition of packaged information resource provision in traditional models of distance learning, which likewise serves to limit learners' engagement with the wider information environment, and the future starts to look decidedly bleak. It has been suggested that unless information services play a central role in educational design and delivery in the future, students on distance and open learning courses which exploit ICTs may continue to be presented with tightly constrained models of learning.[3]

An altogether brighter scenario is suggested by the impact of educational philosophies based on the principles of active learning.[4] Models of networked learning based on these seek to exploit the potential of ICTs to support co-operation and group-work, emphasise the importance of learning in contexts which retain complexity, and incorporate activities to help students develop the process awareness and skills that are essential to independent learning. Information issues are starting to be addressed more often in this context, as it becomes more widely recognised that the ability to use the information resources of the hybrid library effectively will be a key dimension of competence for active learning. From this perspective, courses need to encourage learners to engage independently with the richness and complexity of the evolving information landscape and *build in* – rather than 'bolt-on'– support for developing the relevant skills. Linking skills development closely to subject domain learning is a central tenet here, on the basis that learning is most effective when it involves authentic problem-solving in a meaningful context; reflecting this perspective, recent policy at a national level[5] favours embedding key skills in the subject curriculum rather than tackling them through parallel courses. Equally importantly, the complexity and volatility of the information landscape suggests that relevant knowledge and skills will need to be addressed as a developmental process over time.

Networked Learner Support

Information services in higher education are responding to the explosion in electronic information resources, and to the widespread awareness-raising and skills needs within their user communities, by stepping up their teaching activities[6] and by using ICTs to complement and extend traditional methods of information skills training, user education and reference assistance. Innovations include web-based guides and skills tutorials, online enquiry services using email and the web, mixed-mode or fully online courses in electronic information use, and staff participation in online forums to assist with student project work. There is experimentation with synchronous technologies such as video-conferencing and MOOs for remote reference services, and new initiatives are exploring the potential of applications to enable information staff to 'take control' of users' database sessions as a means of guiding the search process.

At the same time, there is increasing emphasis on academic liaison, to promote the use of electronic resources in teaching and to link skills development activities more closely with the curriculum – something that has long been recognised as important by academic librarians. On the whole, however, librarians traditionally have had little connection with the educational process other than through student enquiries related to assignments, and academics have not necessarily seen the relevance of closer partnerships with them, or have been interested in integrating a focus on information skills into their courses. As they begin to incorporate network capabilities and resources into their courses there is a fresh opportunity for information staff to encourage co-operation and as a result new partnerships are being forged, typically through the development of web resources tailored to specific courses, student groups or subject disciplines.

Collaboration will be a cornerstone of new educational practice for information specialists. If they are to become champions of independent, active approaches to networked learning, and to assert a role for themselves here, they will need to become partners in educational development, and in the design and delivery of courses, *at the level of subject domain learning*. They will need to gain an understanding of information users' needs in the context of specific learning purposes and processes, and contribute to designing on-line curricula, learning resources, activities and environments, becoming involved, for instance, in:

- Raising awareness about information literacy issues amongst teaching staff and others involved in developing online resources and courses, including IT staff and educational developers

- Promoting the value of independent use of information resources in learning, and encourage pedagogic strategies which engage learners fully with the information environment in their subject areas

- Ensuring that information skills development is closely linked to subject domain learning, and advising on ways of putting active learning strategies into practice

- Participating in the design of *contextualised* learning activities for different learning purposes and levels of study – including activities which will be carried out online

- Participating in the selection or design of on-line learning resources

- Advising on the design and exploitation of 'learning environments' and resource management software, and assisting with copyright, metadata and resource-sharing issues

- Providing online tutorial or advisory assistance to support learners' information-seeking activities

- Contributing to course assessment and validation.

Future Directions?

Taking forward this role in networked learner support will be partly a matter of building on the purposes and methods of a well-established tradition of reference and subject librarianship. However, more is at stake than a straightforward extension of traditional practice. Advances in technology, combined with other changes in the educational environment, re-contextualise professional roles in these areas, thereby requiring them to be significantly redefined. The whole professional framework is changing: learner support is becoming multi-disciplinary and collaborative, and assuring the effective use of information resources is part of a broad strategic agenda to contribute to cultural change in teaching and learning. Moreover, changes in the environment have a destabilising effect, calling in to question librarians' 'ownership' of areas related to information access and end-user support: in the shared network space[7] these are not only, or necessarily, their preserve. Familiar divisions of labour between distinct professional groups are under pressure as expertise in technology and information is less easy to differentiate, and as academics create web resources and online learning environments which offer direct access to electronic resources. 'Role convergence' between information and IT staff, and between information and teaching staff, has been identified as a key dimension of the impact of the electronic library[8] and is a source of uncertainty and anxiety for librarians who fear that their traditional expertise may be eroded and their position as intermediaries bypassed.

It is unclear how the roles and responsibilities of the various stakeholders in the educational arena will shake out in the longer term, though it is probably safe to assume that different models of support for networked learning will emerge to suit different institutional circumstances. Still, there is cause for optimism for information professionals who can both adapt to, and shape, the environment as it continues to evolve. Already, the shift of ethos from traditional librarianship to learner support is having a significant impact on organisational structures and individual responsibilities, and staff in many institutions are responding to the challenge to re-define professional roles and relationships. As changes in job titles suggest – e.g. from subject librarian to 'learning advisor' – information staff at the learning interface may well define their professional identity primarily in terms of *educational* practice in the not-so-distant future.

If these staff are to adopt the kind of role suggested in this article, they and their senior managers and will need to engage critically with pedagogic issues associated with new technologies, and participate in institutional debates and strategic planning related to networked learning. Staff will need opportunities to explore new perspectives on their professional roles, to become confident in relation to educational theory and the use of new educational technology, and to extend management and personal skills to multi-disciplinary teamwork and change agency. This in turn sets a new agenda for professional development providers and educators in the information field, whose programmes must be able to meet the needs of professionals whose activities are increasingly defined by the challenge to exploit the Internet as a positive environment for lifelong learning.

References

1. Fowell, S. P. and Levy, P. 'Developing a new professional practice: a model for networked learner support in higher education.' *Journal of Documentation* 51 (3), 1995, pp. 271-280. http://www.aslib.co.uk/jdoc/1995/sep/4.html

2. Noble, D.F. 'Digital diploma mills: the automation of higher education.' *First Monday* 3 (1), 1998. http://www.firstmonday.dk/issues/issue3_1/noble/index.html Last visited 28th April 1999.

3. Stephens, K. and Unwin, L. 'The heart of the matter: libraries, distance education and independent thinking.' *Journal of Library Services for Distance Education* 1 (1), 1997 http://www.wetga.edu/library/jlsde/ Last visited 2nd May 1999.

4. Grabinger, S.R. and Dunlap, J.C. 'Rich environments for active learning: a definition.' *ALT-J, Journal of the Association for Learning Technology* 3 (2), 1995, pp. 5-34.

5. Edwards, C., et al. (eds). *Monitoring Organisational and Cultural Change: the Impact of People on Electronic Libraries. Report of the IMPEL2 Project.* Library Information Technology Centre, London, 1998.

6. Dearing Report. *Higher Education in the Learning Society, Report of the National Committee of Inquiry into Higher Education.* NCIHE Publications (HMSO), London, 1997.

7. Dempsey, L. 'A shared network space: some implications for libraries.' Paper presented at 'E-lucidate: the electronic library for the millenium,' Library Association Career Development Group Conference, Leeds, 14th – 16th May, 1999.

8. Day, J.M. et al. 'The culture of convergence.' *International Journal of Electronic Library Research* 1 (1), 1997, pp. 43-62.

The perspective in this article owes much to collaborative teaching and research carried out with Dr. Sue Fowell of the University of New South Wales, Australia.

25

Beyond the Interface: The Future of Library and Information Services

Carol C. Kuhlthau

What is the future of information? This is an intriguing question for those of us working in the information professions and a particularly important problem for those of us educating future information professionals. Last spring, I participated in a symposium of the Rutgers University Libraries to address the question of the future of the academic library. At this conference the somewhat startling statement was made that we are only at the beginning of the information age rather than in the midst or at the end of this time of radical transition. Technology is developing so rapidly that it is difficult to project the future with any sense of confidence. However, developments in the past five years have revealed the transforming power of emerging information technologies. Two examples of dramatic change are the increasing pervasiveness of the Internet and the rapid development of digital libraries. These are not merely passing trends but signal significant changes in the way we search, access and use information. The vast increase in the amount of information available to a broad section of the global population influences all aspects of our lives. Information is no longer associated with a particular place or specific organisation, but is directly accessible to individual users through advanced computer networks. What is the future of library and information services in light of the dramatic changes in the availability and access of information? This chapter explores some issues related to the future of library and information services in the information age.

User-Centred Services

'User-centered' is the key concept for library and information services in the information age. This idea is being picked up by entrepreneurs in related commercial enterprises, such as Amazon.com, Netscape, and Yahoo!, with the specific goal of providing 'just for you' and 'just in time' service. Traditional library services were established to provide access to selected sources for particular groups of users; however, technology changes the underlying concept of services for users in significant ways. Where the traditional library's emphasis on access provided an important service in an era of limited sources of information, technology offers unlimited sources of information directly to the user. Technology brings the library to the user, rather than requiring the user to go to the library. The library is where you are, ready to address what you are working on, in your time frame. This is very different from the traditional library with its contained collection, catalogued for access, with reference service and bibliographic instruction available for those who came to a designated place at an appointed time. The idea of 'just for you' and 'just in time' service is the key to success in future library and information services as it is in commercial enterprises.

Future library and information services will be called to a mission that goes beyond the interface of the library and information system. The interface, where the user meets the system whether it be the computer screen or reference desk, has been the main concentration of services in the past. User-centered services now call us to think beyond the interface to meet people where they are, to address the tasks that they are engaged in within the context of their lives. User-centered services do not end at locating some relevant sources but have the further role of enabling people to accomplish the tasks that underlie their information seeking. User-centered services sustain and support the information user throughout the information search process. The users' objective becomes the goal of the library and information professional and the service is completed only when the users' task is accomplished. User-centered services have the mission of enabling people to seek meaning in an information-rich environment. User-centered services go beyond the interface with the information system, whether it be a library, a database, or the vast range of the Internet, to address the task of the user.

Evaluation of user-centered library and information services requires judging task completion and impact on people's lives. Evaluation of library services has developed emphasizing a sequence of input, output and outcome. Early attempts to assess the value of services emphasized input, such as the size and content of the collection. Later, measures were developed that judged the value of a service by its output, such as the number of materials circulated and the number of people using the service. Evaluation emphasizing outcome was a major advance in valuing services by addressing such issues as user satisfaction. However, in the future evaluation of services will require ways to determine and describe impact on people's lives in the tasks they are striving to accomplish.

Conceptual Framework of User-Centred Services

The conceptual framework of library and information services shifts when we turn our attention beyond access and location to users' thought processes, information need, meaning making, construction, and use. Bruner's notion of going 'beyond the information given' is beneficial for understanding the act of using information. Using information is more than finding facts and reproducing texts. It involves interpretation brought about through reflection to arrive at a personal construction. In recent studies of information use in the workplace, I have found that an important objective of information seeking is to get a 'new angle' on a problem. People are not just gathering facts but are actively interpreting information and adding value through their interpretations. In these studies, participants explained that they are not just locating the right answer, that everyone else can access, but looking to provide a new way of looking at something that is value-added information. As one participant stated, 'My job isn't to be right all of the time. My job is to help people to reach intelligent decisions on a consistent basis and regularly enough that they will make money.' Insight that adds value to the enterprise is the result of the process of construction in the information search process. Therefore the primary task of information seeking and use is to seek meaning and the goal of library and information services is to enable seeking meaning rather than just seeking information.

For a number of years I have been studying the user's perspective of the process of information seeking and use. Three important concepts that have emerged from this research – process, complexity, and uncertainty – have important implications for user-centered services. An examination of these concepts and how they are interrelated in the process of information seeking and use indicates some important considerations that underlying library and information services that go beyond the interface to address people's tasks and goals.

Process

A model of the Information Search Process has been developed from a series of studies that describe stages of seeking and using information directed toward accomplishing a complex task where significant learning is required. These stages are: Initiation, Selection, Exploration, Formulation, Collection, and Presentation. The process begins with Initiation, when a task is first introduced that requires extensive new information, and moves to Selection, when a general topic or subject is chosen. The stages of Exploration and Formulation that follow have been found to be the most difficult and most misunderstood in the process. Many people expect to move directly from Initiation and Selection into

Collection and Presentation completing the task in the final two stages of the Information Search Process. Exploration and Formulation are the thinking and learning stages of the process when the preparatory work is done that lays the foundation for developing a new perspective. During these stages, a focus is formulated that enables the person to distinguish the relevant from the only somewhat relevant in the Collection stage. The focus enables the creation of a personal perspective or 'new angle' to share or apply in the Presentation stage. User-centered services are based on the user's stage in the process as well as the user's task in incidents of information seeking.

Complexity

People are more likely to experience the stages described in the model of the Information Search Process in some incidents of information seeking more than in others and the degree of complexity of the task that prompts the information seeking seems to have a direct relation to the user's experience of the information seeking process. Three degrees of complexity were identified in a recent study of information use for tasks in the workplace. Tasks that required routine monitoring, such as tracking a trend, were considered to have low complexity. Tasks that required one source of information, such as answering a specific question, were considered to have moderate complexity. Tasks that required extensive new learning, such as initiating a new project, were considered to have high complexity. A complex task is defined as such by the user and is not necessarily complex in and of itself or complex for everyone. Complex tasks were generally those about which people knew little in the beginning, had to learn about through information seeking, and had to act on their new learning by producing a report, a product or other significant action. When asked to describe different tasks that require information, participants differentiate between routine and complex tasks and clearly describe a different experience in the process of seeking and using information in more complex tasks. Routine tasks, in which they apply familiar patterns of information seeking and use, were experienced with confidence and ease. Complex tasks, in contrast, require substantial new learning and were associated with considerable apprehension and uncertainty. These initial findings indicate that complexity and the experience of uncertainty depend on the knowledge state of the user. Tasks cannot generally be classified as routine or complex for all people in all situations. User centered services take into account the knowledge state of particular users in order to provide 'just for you' services.

Uncertainty

Uncertainty commonly increases in the Exploration stage of the Information Search Process. Library and information professionals providing user centered services need to be alert for signs of uncertainty and find ways to accommodate users in the course of their uncertainty and apprehension in information seeking. A principle for library and information services is proposed that incorporates the concept of uncertainty. Uncertainty is a cognitive state that commonly causes affective symptoms of anxiety and lack of confidence. Uncertainty and anxiety can be expected in the early stages of the Information Search Process. The affective symptoms of uncertainty, confusion, and frustration are associated with vague, unclear thoughts about a topic or question. As knowledge states shift to more clearly focused thoughts, a parallel shift occurs in feelings of increased confidence. Uncertainty due to a lack of understanding, a gap in meaning, or a limited construct initiates the process of information seeking.

At the beginning of a complex task when persons have little prior knowledge about the subject, they encounter much information that is unfamiliar, novel and unique. This incompatible information is inconsistent with what they already know and is not easily brought together into a conceptual whole. The experience of uncertainty is likely to intensify at the very point when the person is expecting to have the work come together for him or her. This experience is common in the Exploration stage when ideas need to be mulled over and reflected on to form an understanding or personal perspective in the Formulation stage. So in a complex task, information is initially likely to increase uncertainty rather than reducing it. A conceptualisation of information seeking and use as a constructive process, where uncertainty is the beginning of accomplishing a complex task, underlies user-centered library and information services.

The Future Role of Library and Information Services

In my book *Seeking Meaning: A Process Approach to Library and Information Services* I describe five levels of intervention in library and information services. The levels of intervention are delineated within the two traditional library services of reference and bibliographic instruction to provide a way of envisioning how user-centered services might be implemented. Reference services are differentiated in five levels of mediation: organiser, locator, identifier, advisor, and collaborator. Instructional services are differentiated in five levels of education: organiser, lecturer, instructor, tutor, and counselor.

The organiser at the first level provides the foundation for access to information. A major contribution of librarianship is the systems for classifying, cataloging, indexing, and abstracting that efficiently provide access to collections of materials. This unique contribution will continue to be important to accommodate new information technologies. The organiser role offers the prospect of bringing some order to the vast unorganised information produced by emerging technologies. New systems will need to be developed that are tailored to particular groups of users as well as systems that work for general audiences.

The next three levels of intervention, the locator/lecturer, identifier/instructor, and advisor/tutor, are common in traditional library services of reference and bibliographic instruction. The locator intervenes with one source in a single reference encounter and the lecturer provides instruction in a single session for general orientation. The identifier intervenes with a group of relevant sources and the instructor teaches about these sources in one independent session. The advisor recommends a sequence to use relevant sources and the tutor instructs on a strategy for navigating through a search for relevant sources. Although these services still have value they diminish in importance as individual users perform many of their searching and retrieval tasks on their own. These levels address the traditional role of the service centering on location and access.

At level 5 the collaborator/counselor addresses the holistic experience of seeking meaning within the process of information seeking. This role accommodates the constructive process of users and needs to be developed for future user centered services.

Library and information services in the future will centre on the organiser role at level 1 and the collaborator/counselor role at level 5. The organizer will develop systems of classification and cataloguing for organizing the vast resources of the technology. The counselor and collaborator will support the user in the constructive process of information seeking and in accomplishing a complex task. Counselors advise in the stages of the Information Search Process and instruct in information literacy. Collaborators work as partners in accomplishing tasks requiring extensive information.

There is a need to redefine the roles associated with information provision to be more interactive, collaborative. The collaborative role may require the librarian or information professional to enter into a partnership with the user to accomplish the information-seeking task. In this partnership the librarian has expertise of resources and processes whereas the user brings knowledge of content and context. The aspects of information seeking and use with which users need help are in the ongoing thinking process related to interpreting and connecting the disparate pieces of information gathered in order to provide value-added information. This is new territory for the librarian whose traditional role of providing access to relevant references and sources has stopped short of the process of making meaning. A collaborator, however, is called upon to work on the same task as the user but in a different capacity and with different talents.

There is a need to develop ways to diagnose a zone of intervention to respond to uncertainty, complexity and process within complex tasks.

In a similar way, the counselor role is engaged in information literacy rather than library skills. Traditionally the librarian has been involved in teaching library skills and more recently information skills. But future services will require librarians to have a broader view of their role in instruction to incorporate the promotion of information literacy, that is, the ability to use information intelligently in all aspects of daily living. Information literacy prepares people for the complexity of seeking meaning in the information rich environments of new and emerging information technologies.

The concept of a zone of intervention has been introduced to enable library and information professionals to provide user-centered services by diagnosing users' needs for opportunities for effective and efficient intervention. The zone of intervention is that area in which an information user can do with advice and assistance what he or she cannot do alone or can do only with great difficulty. Intervention within this zone enables the user to move along in the information search process. Intervention outside this zone is inefficient and unnecessary, experienced by users as intrusive on the one side or overwhelming on the other.

When we look beyond the interface, the goal of user-centered information services shifts from seeking information in traditional library services to seeking meaning in the process of construction. Library and information services have two essential roles in the information rich environments, the role of organiser of diverse, rapidly-produced information and the role of collaborator/counselor in using information for the tasks of daily living.

Education for User-Centred Information Services

Educating library and information professionals for the future will encompass more than imparting a set of skills. Future library and information professionals will need to create new services for a changing information environment. Students will need to understand the traditions of classification, cataloging, indexing, and abstracting and be able to adapt these systems and create new systems for the evolving information environment. Students will need to understand the constructive process of information seeking and be able to recognize a zone of intervention to provide counseling and collaborating that enable users to accomplish their desired extensive tasks.

Designing User-centered Library and Information Services is a new M.L.S. course at Rutgers University School of Communication, Information and Library Studies that strives to prepare students for serving in new information environments. The students identify a group of people with a significant task called a community of practice. They examine the nature of that task, what sources of information are used, what strategies are applied and how effective the sources and strategies are for accomplishing the task. The assignment is to design a library and information service to improve on the sources and strategies they are now using. Students are developing understanding, knowledge, and ability in designing 'just for you' and 'just in time' service that is personalised and tailored to the specific task of this community. In their investigation of a community of practice they find that people often attempt to create their own database or files of information that specifically addresses their task and that call for the expertise of the organiser role that the library and information service can provide. In addition, they find that people have difficulty putting ideas together to accomplish a task and call for the collaborator role that the library and information professional can provide. They also find that people need help in learning how to access and use information effectively for a range of purposes and need the information literacy that the counselor role can provide.

Patterns in users' tasks in different environments suggest that information use in the next decade and beyond may call for library and information services that enable the process of meaning making of information users. Future services will need to sustain people through substantial construction in the process of information seeking in order to provide value-added information to accomplish extensive tasks.

26

The Three Properties of Information: Content – Structure – Publication

Gerry McGovern

What is Information?

Today, information is everywhere. We need it to work. We need it when we buy things. More and more of what we create and what we buy is made up of information.

Information is indeed everywhere and if we want to play an active part in the Digital Age, we must become 'information literate.' This requires much more than reading and writing; it requires a deep understanding of what information is and how it behaves within the Internet environment.

So what is information? Many would see information as a type of commodity or resource, as something solid, something identifiable. If information is indeed a resource or commodity then it certainly does not have the same characteristics as traditional resources and commodities. Resources tend to have a limited supply. Some are renewable such as fish and forests, though they take time to renew and can be depleted if not properly managed. Some, such as oil and gold are not and are of finite quantity. Commodities have a definite cost attached to their creation, and to the creation of each new unit in any particular line.

Information behaves differently:

- Information cannot be consumed; it can only be shared
- Reproduction of information is generally cost effective

When information enters the digital realm, the above attributes are even further enhanced. The Internet allows information to be shared all over the world. While it is relatively expensive to create a physical copy of a book or article, there is an almost zero cost to copy a piece of digital information.

So if information is not really a commodity or a resource, what is it? Part of the answer can be found in exploring the very meaning of the word 'information.'

The Merriam Webster dictionary defines 'information' as 'the communication or reception of knowledge or intelligence.' The root of 'information' is 'inform' which itself comes from 'form'. The word 'form' finds its origin in the Latin 'forma', which means 'shape'.

Merriam defines 'inform' as 'to give character or essence to: to be the characteristic quality of: to communicate knowledge to: to impart information or knowledge.' It defines 'form' as 'the shape and structure of something as distinguished from its material: the essential nature of a thing as distinguished from its matter.'

Chambers Dictionary defines 'form' as a 'shape: a mould: something that holds, shapes: a species: a pattern: a mode of being: a mode of arrangement: order: regularity: system, as of government: beauty.' It defines 'inform' as 'to give form to: to animate or give life to: to impart a quality to: to impart knowledge to.' It defines 'information' as 'intelligence given: knowledge.'

Webster's Dictionary, 1913 edition, defines 'form' as, 'The shape and structure of anything, as distinguished from the material of which it is composed; particular disposition or arrangement of matter, giving it individuality or distinctive character; configuration; figure; external appearance.' It defines 'inform' as, 'To give form or shape to; to give vital organizing power to; to give life to; to imbue and actuate with vitality; to animate; to mold; to figure; to fashion.' It defines 'information' as, 'The act of informing, or communicating knowledge or intelligence.'

From the above discussion and definitions, we can isolate the primary characteristics that true information must exhibit. These are:

1) Information is a process or activity, and not an object
2) Information communicates knowledge and intelligence
3) Information is made up of three essential properties: structure, content and the communication or publication of that structured content.

The Three Properties of Information

The three properties of information are:

1) Content
2) Structure
3) Publication

Content is the message. It's what you're trying to say. Content is the starting point of all information but it is *only* the starting point.

Structure is how content is put together – it is the 'form' in 'information'. Without structure, content has very little value as it cannot be easily communicated. Structure begins with how a sentence is fashioned. It moves to a paragraph, chapter, table of contents, index, etc. On a web site, structure is how you organise and link your information. Lack of structure is behind much of the information overload on the Internet.

Publication is the final property of information. Information, as we have seen, is a verb, not a noun. It is a process, not an object. It is the *'act* of communicating knowledge or intelligence.' Publishing is the act of making something public, of getting the structured content out there to your target market. If you don't publish your information, then it doesn't really matter how good the content is or how well it is structured, it will not reach your audience and therefore you will have wasted your time.

It is important to note here that creating a web site is *not* necessarily an act of publication. The fact that you put a web site on the World Wide Web is like opening a shop in your bedroom, or a store on the North Pole. Sitting deep among the millions of other web sites, practically nobody will know that it exists unless you actively publicise it.

Measuring Information Value

Webster's dictionary defines value as, 'a fair return or equivalent in goods, services, or money for something exchanged'. Therefore, we could say that information value can be defined as, 'a fair return or equivalent in goods, services, or money for information exchanged'. To measure the particular value of a piece of information, using the three properties of information, gives us the following formula:

$$\text{Information Value} = \text{Content} \times \text{Structure} \times \text{Publication}$$

Calculating Information Value involves multiplication rather than addition. To illustrate the multiplier affect, let's say you have thirty hours available to create Information Value. If you spend 15 on Content and 15 on Structure and 0 on Publication, what would be your Information Value? It would be $15 \times 15 \times 0 = 0$. Not publishing your structured content would have been like printing up 100,000 magazines and leaving them in your office.

We should look at the above formula as indicative, but as a general rule it would be true to say that the level at which you will achieve maximum Information Value is when you are spending an equal amount of time at each property. So, in the original example, you spend 10 hours working on the content, 10 hours working on the structure and 10 hours working on the publication. Thus, the formula would be: $10 \times 10 \times 10 = 1000$.

The Changing Role of Information

It has been widely stated that one of the prime reasons that accentuated the 1998 Asian economic meltdown was the lack of proper information flows in Asian economies. The *New York Times* in October 1998 stated that a major report on the global economic crisis had, 'called for the release of far more economic data by countries and companies alike, noting that the lack of such information "exacerbated" the panic that began in Asia last year.'

The *New York Times* also wrote about a World Bank report, which had noted that, 'information gaps had contributed to the Asian financial crisis. It found that the capital outflows and currency collapse experienced in Thailand, South Korea and Indonesia "reflected the pervasive lack of information throughout the world about finance in the region."'

Whether we like it or not, we live in a world where *the flow* of information is becoming more and more important. It doesn't matter how well a company is actually performing, if it doesn't properly inform the market and the world, then its stocks will tumble.

Historically, it was not so much information that was seen as power, but actually the *control* of information. Power was wielded by the information you did not give, and many governments, businesses and religious institutions operated with this as their primary principle. Information was seen as gold, and hoarding and protecting it was a favourite activity.

It was also true to say that in an Industrial Age society the vast majority of people did not require much ongoing information to do their jobs. Much work either required physical labour and skill (such 'information' being acquired by apprenticeship) or it was administratively repetitious.

Formal education, for the vast majority of people, was not therefore about learning knowledge but rather about learning the basics (reading, writing) and most importantly, 'learning your place.'

In fact, for those in power, there was no incentive to allow information to flow throughout society because this was likely to raise people's awareness and expectations, throwing a light on their monotonous lives and poor living conditions, thus quite probably causing social unrest.

In the Digital Age, the function of information has changed as a result of:

- Modern machinery and computers continuing to replace humans in the sphere of physical labour and skilled trades
- The explosion in cheap, far-reaching information tools: printing, radio, television, Internet
- The explosion in the need for information from a growing Digital Age knowledge workforce

While in the Industrial Age information was like gold and hoarded, in the Digital Age it is like milk and needs to flow quickly to a hungry knowledge workforce, or to a consumer society hungry for news (and scandal). Today, a society that does not allow for the rapid diffusion of information at every level is quite simply not an Information Society, and is merely shoving its arm in the dyke of the modern media-saturated age.

So much about the Digital Age makes it difficult to hoard information. The very act of organising and storing information is becoming an increasingly difficult task. In this fluid, digital world, trying to keep such information 'safe' and out of the flow is equally time consuming. Companies are finding that with the Internet and other networks it is almost impossible to lock information away.

In many circumstances, the time spent keeping information safe would have been better spent acting upon the information. And of course, you may find that no sooner have you organised and made the information 'safe' than it has lost its value – gone sour.

Information Overload

Industrial Age societies often faced issues of scarcity; scarcity of oil, scarcity of fish, scarcity of timber. Many Industrial Age scarcity problems were as a result of careless management of resources. We have the same careless management of information and content today, only this time we are creating a *glut* of information, rather than a scarcity. One is as bad as the other.

Think about it for a moment and ask yourself the following question: How well do I manage the space on my hard disk? I know what my answer is: 'Not very well.' Every day, drafts and drafts of digital documents are produced. Files are stored, not because of their value, but because it is easier to store them than to delete them. Things are published on the Internet because they *can* be published. Emails are sent because it is easy to send them, not because of their communication value.

Information overload is the single greatest problem not simply facing the Internet but facing all of us in practically every aspect of life in the Digital Age.

Information overload can manifest itself in a variety of ways:

- Lack of planning with regard to how information is organised leads to an environment where it becomes increasingly difficult to find what you're looking for
- Lack of 'weeding' of old and redundant information obstructs the paths to quality information

- Lack of proper standards with regard to the publishing of information results in poor quality and error-prone content
- The ease of copying and cheap storage encourages people to create masses of generally unnecessary copies
- People can receive so much 'quality' information that they simply do not have time to take it all in
- People who have not been trained in information analysis and management feel overwhelmed even when they have to deal with what should be average information demands
- People can become 'addicted' to information, always wanting more information before they make a decision, so that they rarely make the decision on time

The signs of information overload are all around us today. A 1998 study entitled *Workplace Communications in the 21st Century*, conducted by the Institute of the Future for Pitney Bowes, had some interesting findings. 'The average worker across a broad range of positions, from administrative to senior executives, say they now send or receive approximately 190 messages on any given day,' it stated. 'This volume of messaging, and the corresponding demands of managing the flow and responding in a timely and efficient manner, now shape how people in many different positions and industries actually structure their day.'

A 1997 Reuters survey of 1000 business managers found that a growing number of them were addicted to information and the Internet. More than 50 percent said that they didn't have the capacity to assimilate all the information they were getting. 97 percent believed that their companies should provide courses in information management training.

In April 1998, NEC Research published a survey which claimed that there were already 320 million pages on the Internet and that this would grow by 1000 percent over the coming years. The research claimed that even the most comprehensive search engine was only managing to categorise 34 percent of all available pages. 'Hundreds of pages are being added constantly,' survey co-author Steve Lawrence told the *San Jose Mercury*. 'There is no simple way to index it all. There could be any percentage of pages out there that nobody has actually accessed yet.'

The Need for Information Infrastructure

New search technologies may make it easier for people to find what they are looking for, but it is my firm belief that information overload is a more fundamental problem that will not be solved by new search engines alone. Companies and countries and other entities that want to gain proper benefit from the Internet must plan for and install a comprehensive information infrastructure.

The key challenge today is to order what we have and not to create more information waste. A city that does not have an overall plan that covers housing, traffic, green spaces and amenities, etc., is one that invites long-term chaos and decline. Seriously developing for the Internet requires a comprehensive information plan. Otherwise, poor content and poor structures will choke off much of the useful information. Poor information will grow like weeds.

An organisation will require a sound information architecture foundation if it is to build anything that will have a long-term capacity to grow and remain useful. Organisations will require information planners, editors, archivists and 'librarians,' whose job will be to put order on the information, to place the information in context, on its proper 'shelf.'

I don't know who described the Internet (circa 1995-99) as like a library with all the books on the floor and with the lights turned out, but it was a fairly apt description. We need to turn the lights on and start putting the 'books' onto their proper shelves.

In 1998, it became generally recognised that information overload was reaching a crisis point. Inventor of the World Wide Web, Tim Berners-Lee told the *Los Angeles Times* that 'Whereas phase one of the web put all the accessible information into one huge book, if you like, in phase two we will turn all the accessible data into one huge database.'

'This will have a tremendous effect on e-commerce,' he continued. 'You could say, "Find me a company selling a half-inch bulb to these specifications," and a program will go through all the catalogs – which may be presented in very different formats – and figure out which fields are equivalent and then build a database and do a comparison very quickly. Then it will just go ahead and order it. It would be a real mistake for anyone to think the Internet is done or the World Wide Web is done. We're just at the start of these technologies.'

Indeed we are just at the start of a very long road. Creating a widely accepted comprehensive information infrastructure on the Internet is a massive task that will take years to complete. Some preliminary steps on this road have been taken. The 'Dublin Core' initiative and 'meta tagging' being two examples. However, so much more is still to be done. A broadly accepted international standard, such as the Dewey system for libraries, which would deal with the organisation and categorisation of information, is badly needed.

At a basic level, this information infrastructure will have to categorise information *before* it is entered onto the Internet, not after as is the case with search engines and directories today.

Think of it this way. Most books published today have some sort of library categorisation information. On the book's spine and back cover is a unique number, as well as a subject categorisation (business, philosophy, fiction, etc.). Business books, for example, are rarely published without a table of contents, with many books having indexes as well.

In the future, every picture or image, for example, stored in cyberspace will be asked a certain number of questions: When was this picture taken? Who is the copyright owner? Where was it taken? What geographic category does this picture fit into? What information category does this picture fit into? An equivalent set of questions will be asked of an article or other unit of text. Once these questions have been answered, it will be possible to build a proper map which will allow people to find what they are looking for in the most efficient way.

27

Helping Small Business Encounter Information

Sheila Webber

The Future of Information?

There will be a lot of it about. People will interact with it in a disorganised way (although many of them will think that their information habits are OK, insofar as they think about them at all). Most information will not be managed by trained information professionals. The greater part of it will disintegrate or be discarded, and thus be lost to posterity.

So, no change there then.

My intention is not to be defeatist, but to put the issue of information management in perspective. There is more written about the way in which information is used in organisations and libraries, than about the way in which *individuals* interact with information. However, there are many more individuals than there are organisations, and (after all) organisations are made up of individuals. In particular, small businesses may have information habits which are more like those of individuals[1]: and there are many more small businesses than large ones. Small businesses seem to rely more on personal contacts, and less on formal information channels. This is scarcely surprising, since small businesses by definition consist of relatively few people.

Looking at the information activities that people engage in through the centuries, one sees the same sort of things happening: people deliberately destroying or preserving cultural heritage; variously stealing, losing, organising, concealing or flaunting corporate information; quarrelling over intellectual property rights; hoarding or burning letters; seeking to ensure that their own version of history is the one to survive.

The basic functions are much the same, but the way in which technology helps or hinders these activities is constantly evolving. Of all the PEST factors (looking at Political, Economic, Social and Technical issues), technology seems like the wild card. History tells us that the ideology of those in power will have a big effect on what is kept, broadcast or forgotten. Information activity will be influenced by the economic health of a country, organisation, or family. People's habits and tastes will affect what is published and preserved, whether these tastes reflect local culture or global fashion trends.

But technology will also influence things, and sometimes in unpredictable ways. For example, email seems to be strengthening both local and global contacts. Teenagers can keep up with schoolfriends easily once they move to university. Writing a letter requires effort, whereas sending email is an enjoyable displacement activity. At the same time, you may already have a circle of friends in other countries who share your passion for gaming, or saving the earth, or Madonna.

Potentially, the Internet offers a huge archive of diverse information, a goldmine for future social scientists. But many web sites are already lost for ever, whilst academics and librarians puzzle out what and how to archive. The Internet provides a rich (if biased) snapshot of life this minute, but only a partial glimpse of life as it was last year.

Another effect of the Internet is to make more individuals and small organisations potential information providers. Information formerly hoarded in shoeboxes or locked in elderly computers can now be put on web sites, ready for anyone to tap into. Along with the mundane and the offensive there is valuable information. When one expert mounts information, other enthusiasts may emerge from around the world, and offer their contribution to the site. The word 'communities' is over-used in this context, but there certainly are sites where people with common interests congregate and sometimes contribute. It seems likely that there will be more few-to-few information transactions, in contrast with the one-to-many model familiar from mass media, or the one-to-one model that characterises phone based or face-to-face information sourcing.

But are people getting any better at finding information? And can information professionals hope to provide a better service to small businesses in the age of the Internet? Various studies have highlighted the fact that businesspeople do not see information seeking as a separate skill, and that they see 'business problems' not 'information problems'.[2] This has meant that businesspeople may be difficult to convince that a librarian has anything to offer them: if all they perceive is a business problem (e.g. finding new customers) it will not be immediately obvious to them why a librarian, who knows little about their business, should be able to help. If someone does not think they need information and does not think they need help or training in gathering it, then presenting a logical argument about the value of formal information services and people will have no impact on them.

Repeatedly proving value, by presenting the right information at the right time in the right way could win people over. However, doing this reliably for individuals and small businesses is often prohibitively expensive. If people are not good at recognising their own information needs, then the service has to be proactive (but not perceived as intrusive). It has to provide the right information: difficult for small businesses with a very precise, local market. Business information services promising 'information at your finger tips', but failing to deliver on specific enquiries (because it turns out, for example, that no published sources have much to say about the market in Ardnamurchan for hand-made duvet covers) are not going to do much to advance the cause of the information professional.

Similarly counterproductive is the hype saying that information is easy to find on the Internet. This is, of course, nonsense. It is true to say that, once you have access to the Internet, finding *particular types* of information becomes easy, and is going to get even easier. Finding recent quotes for shares on major stock exchanges, or the latest local headlines is easy to do, because they are being funnelled into all sorts of frequently used sites. However, the maker of duvet covers setting up business in Ardnamurchan might still be at a loss. To find relevant information unaided, the manufacturer would have to develop a sophisticated search strategy, or trust to luck that eventually he or she might stumble across something relevant in their ramble from site to site.

Access to the Internet seems likely to change people's information-sourcing habits. Ease of use is often identified as a key factor in preferences for information gathering. Studies are already beginning to indicate that people are finding it more convenient to find something on the Internet. Sitting at a computer and clicking a few keys can be less costly, in terms of time and bother, than making a trip to the library, or picking up the phone and verbally negotiating your way round an organisation to find the information you want.

Therefore, in the future, businesses may use libraries less (or rather, even less, since small businesses are not great users of libraries anyway). They may prefer online directories to print directories, once they are convinced of their reliability. They may turn to their local virtual community, rather than their local Chamber.

Whilst they may end up using a slightly wider variety of information sources, in the end, though, they are likely to be accessing the same *types* of information (on suppliers, markets, loans, health and safety and so forth) but through different channels. Internal information, and information shared with trading partners, is more likely to be electronic, and more amenable to storage, sharing, organising and manipulation. One hopes that this means that people will make better use of it. However, people whose natural disposition is to gather information by browsing or by bumping into it are not necessarily going to become ace users of search tools, just because there are more of them around.

For this reason, research looking at how individuals gather information is fascinating. For example, Sanda Erdelez has investigated people 'bumping into' information: what she calls 'information encountering'[3]. She classifies people into 'non-encounterers', 'occasional encounterers', 'encounterers' and 'super-encounterers'.

At one end of the spectrum are the super-encounterers, for whom life is one big sea of information. Whatever the context, if they bump into a piece of information that they think is of value to them, to their colleagues or to their family, the odds are that they will gather it. The fact that, at that moment, they are not in an 'information' environment, but (for example) in a launderette idly turning over other people's rejected reading matter, does not deter them. If they spot something interesting, they will copy or pocket it, and later post it, file it or pass it on.

At the other end of the spectrum are those unfortunates who tend only to acquire information when they are consciously on the search for it. Even then, they will only be aware of the information on the topic that they want at that specific moment. I say 'unfortunates', since I immediately identified myself as a super-encounterer: one who tears articles out of in-flight magazines, who emails web addresses to colleagues, and who returns from a trip to the library with a pile of photocopied material about a range of topics, even if the original intention was to look up one specific article.

However, the term 'unfortunates' reflects my prejudices. Really, one of the key effects of reading about Erdelez's research was to make me understand more about the other person's point of view – or rather, that there *was* another point of view; that this was an interesting way to classify people's attitude towards information which could be a help in meeting people's information needs.

There is a continuing need for information professionals, but they have to expand their idea of what information is, and spend even more time studying what individuals do outside the library or information centre. This includes puzzling out strategies to help people who:

- are non-encounterers who do not gather information until they have recognised an information need, and
- find it hard to *recognise* an information need in the first place.

Theories and systems often seem based on the idea that people are aware of the value of information, and are purposeful information-seeking creatures, happy to exchange information with one another for the common good. These theories and systems are surely doomed to failure, or at least only partial success.

It is important not just to create tools for the active information-searchers, but also to create environments that suit the browser and encourage information encounters. The web seems an obvious browsing environment, but it is not perfect. There are lessons to be learnt from the way people interact with libraries, and with the other information channels that they use.

Trying to understand people's motivations and habits is vital. Analysis of these habits can guide the design of pathways and signposts through the electronic web of information. The encouraging thing for job-seeking information professionals is that a human is likely to be better at this than a machine, and more sensitive to changes in need and preference.

This is obvious, but the big research money seems, too often, to go into developing the technology, and investigating how people search (rather than how they browse or amble across information). There is good research into browsing behaviour,[4] but more effort is going into tweaking search engines. One also cannot be too optimistic about web page creators cataloguing and indexing their sites consistently. A study in *Nature* estimates that only a third of web sites have keywords or subject descriptions added to make them more retrievable, and that even the best of the search engines only indexes 16% of pages on the Internet.[5] Some extremely useful information will be amongst that unindexed material. Again, this points to the need for people who can trace and evaluate information, and then guide the user.

In a few-to-few environment, information professionals may actually have more of a chance of helping small business. There is more possibility of tracing specialised sources and local experts via their web sites. There is the possibility of pooling resources with other agencies to create better virtual libraries than one could physical ones. Personal contact is still important. It helps to build trust, and keeps one in touch with user needs. It makes it easier to introduce people to short cuts through the web and highlight places where information is most likely to be encountered.

The Internet is making an impact on information seeking and information use. However, there has not been a paradigm shift in information seeking: people have a lot of the same types of information need and are not going to mutate into super-encountering super-searchers simply because they have a web browser on their desktop.

Information professionals can, though, capitalise on the opportunities that the Internet creates: to find out more about people's needs and habits, to communicate with them effectively, and to shorten and make more pleasant the pathway between the information seeker and the information sought.

References

1. See for example Vaughn, L.Q. 'Information search patterns of business communities: a comparison between small and medium sized businesses.' *Reference and User Services Quarterly* 37(1), 1997, pp. 71-78.

2. See for example Roberts, N. and Clifford, B. 'Regional variations in demand and supply of business information: a study of manufacturing firms.' *International Journal of Information Management* 6(3), 1986, pp. 171-183.

3. Erdelez, Sanda. 'Information encountering: it's more than just bumping into information.' *Bulletin of the American Association for Information Science* 25(3), February/March 1999, pp. 25-29.

4. There is a useful review in Chun We Choo, Brian Detlor and Don Turnbull, 'Information seeking on the web: an integrated model of browsing and searching.' Paper presented at the 1999 ASIS Annual Meeting, October 1999. http://choo.fis.utoronto.ca/fis/respub/asis99/ (visited 2 August 1999).

5. Lawrence, Steve and Giles, C. Lee. 'Accessibility of information on the web.' *Nature.* 400, 8 July 1999, pp. 107-109.

28

Information, Communication, and the E-Generation

Jane Klobas

New Year's Day 1999. The last New Year of the 1900s. The first New Year that my family in Perth, Western Australia, used Internet videoconferencing rather than the telephone to celebrate with friends and family in Italy. The adults were excited! There was no need to book a telephone call, no little *beep*s to count the minutes. New friends in Italy were able to see my husband rather than rely on my description of him. We revelled in the leisure of being able to raise a glass to one another and engage in that desultory conversation that marks post-New Year's Eve gatherings everywhere.

Why, you might be asking, am I telling you this? This essay is not about Italy or friendship. It is, though, about communication, about how our children will be communicating with friends and colleagues, and how they will be gathering and sharing information in the future. So, let me return for a moment to New Year's Day 1999.

Two thirteen year olds were behaving as thirteen year olds do: holding court, expressing opinions, and generally trying the monopolise the communication channel. Internet videoconference? Telephone? *Non c'è differenza*! It's all the same to a teenage girl. The seven year olds were bored after one or two minutes of saying hello to their parents' friends, and ran off to do something much more interesting.

Listening to and watching these children who took international videoconferencing for granted, we couldn't help wondering: What will the world of work be like for these children? How will people who grow up with the Internet (or its successors) gather and share information during their working lives?

Where can we begin to answer these questions? Our New Year 1999 not only raised the questions but gave us a starting point for answering them. Communication is a human trait. People will incorporate effective new media in their communication patterns, but the fundamental nature of human communication will remain the same. The adults were simply using the Internet as a more effective, more readily available, and less expensive medium for their normal New Year's communication. The teenagers were testing their ability to contribute to (to take control of?) adult conversation, and the medium through which they tested this was not a concern. The young children were, as usual, more interested in active play than a discussion that involved the interplay of only words and ideas.

We therefore have a force for change in the way people communicate: the Internet is a new communication medium; but we also have a principle that can help understand the impact of that force: communication is a human trait, and regardless of medium, the fundamental nature of human communication remains the same. We can apply this principle to understand changes in communication of work-related information and knowledge that has been prepared and recorded by others.

How will people gather and share recorded information and knowledge for their work in the future? An obvious first response to this question is: *electronically*. There is now no doubt that information will be distributed electronically, through the digital networks that will evolve from the Internet. It is no longer difficult to imagine the world of the *digital appliance* which can be connected to a distribution network not unlike the household and business appliances that connect to water, power, and telephone networks today. Domestic products that incorporate some element of this thinking are already available: from the digital television that provides Internet connection to more speculative products such as the GSM-connected refrigerator that enables Internet access and control of domestic appliances from a WWW browser in a flat screen on the refrigerator door. It will not be long before information distribution through the digital appliance will be commonplace, straightforward, and as well understood as information distribution through newspapers, journals, and books. (And, no, I do not think electronic information distribution will replace print, but it will be a widely used and readily available medium that will find its place among the print media.)

The big question, as always, will be *'How will people find the recorded information that they are searching for?'* Using the principle established earlier, we can expect that the fundamental nature of information seeking behaviour will remain the same.

How do people locate information now? We know that, for most, the first point of contact is another person. Business people telephone or email a colleague who might know the answer to their question, or walk to a neighbour's office or desk; sometimes, they might even consult a librarian. Academics and researchers rely heavily on referrals from colleagues or writers who they trust as experts in the field of interest; sometimes, when an area is new to them, they consult indexing and abstracting services; sometimes they play pot luck with an Internet search engine; increasingly, they have access to specialist indexing resources on the Internet, bibliographies or web pages that respected colleagues have prepared to point to print or Internet resources they have found useful.

There is no reason to expect the fundamental basis of this communication to change. People will continue to trust other people, to ask them for information, and for references to source of information. Increasingly, they may use electronic means to make these requests: email, discussion databases and *listserv*-like ongoing conferences, live 'chat' connections, electronically communicated video, voice, and text. Regardless of the medium, they – we! – will continue to prefer information and sources of information recommended and endorsed by trusted colleagues and experts.

Who will those trusted colleagues and experts be? As now, there will be in each profession, industry, and organisation, a handful of gatekeepers who play the role of gathering and disseminating useful information to others. There will still be information specialists, who may or may not still be called *librarians*. These information specialists may be self-employed professionals working in loose coalitions of professionals from a variety of fields who form project-related teams, un-form, and re-form again in different configurations for different projects. They may be members of more permanent teams in larger organisations, the information recording and information gathering experts, who are respected for their ability to keep in contact with sources of recorded information, and to sift and reorganise it for the team's work.

There will be other changes. Some educators are already teaching a new form of information literacy to university and school students. It is not difficult to imagine school children being taught many of the 'secrets' of librarianship as part of their literacy education. By the time they leave school, the e-generation may well know not just about decimal classification, but about how classification schemes can be used to provide a framework for recorded knowledge. They will almost certainly know about different ways of ordering references to documents (alphabetical, by class or subject, by geographical location, by time, etc.). They are likely to know the difference between pre-coordinate and post-coordinate indexing, or at least between describing a document's content before storing it or allowing others to use free language to retrieve it at the point of need. They will put this knowledge to good use in their own information gathering. The information professional will therefore be more specialised than now, and will be working with a more knowledgeable clientele.

And someone will still be needed to put indexes, abstracts, 'bibliographies', reference sites, and other organising schemes in place. Members of the information-literate e-generation will be able to describe their own records or documents, using description schemes developed by others. The schemes may perhaps be extended by automatic indexing systems, but the history of automatic indexing and language translation systems tells us that human language is complex and human ideas grow and change so quickly that machine indexing cannot successfully encompass the range of ideas in most complex business and research records (the very records that are most difficult to find). For the same reasons, machine-assisted retrieval will have its limitations, it will always remain machine-assisted retrieval to the human process of interpretation and processing of knowledge and ideas.

What, then, will be the shape of our information world when today's seven year olds join the workforce, say in 2012? Many of us will still be working in the information field. We will have adjusted to the technological and industrial changes that began with widespread availability of public, computer-based communication networks in the late twentieth century. And we will be watching the ease with which members of the e-generation take for granted an information world that we can regard, with the benefit of history, with awe and the pleasure of knowing that we grew with it and perhaps even helped to shape it. Many of the e-generation will have the ease with electronic sources of information that only online experts had in the 1980s and early 1990s. They will understand how knowledge is built, how it is recorded, and how it can be retrieved. And they will know when to call in the experts to build knowledge structures, or to help retrieve information from them.

As we begin to understand what widespread electronic information storage and retrieval might mean, we can watch the behaviour of today's school children to understand that, no matter what the prevailing technology, human communication remains much the same. Teenage girls try to monopolise the communication medium, regardless of its apparent sophistication, and young children are not particularly interested in adult conversation, even if it is occurring live at virtually no cost by video across tens of thousands of kilometres. Some fundamental patterns of communication at work are also unlikely to change. People use the information literacy they gained at school as they deal with information in the business world, and professionals prefer to obtain references from colleagues and experts than to take a time-consuming and frustrating stab in the dark for information.

In sum, technology may change, but people remain the same. The structure of industry will change with changes in technology, and education in electronic information literacy will increase people's knowledge of how information is recorded, indexed, and retrieved. Together, these foundations will lead us into a business information world in which electronic information skills are commonplace, but experts in indexing and specialists in retrieval are especially highly regarded – an exciting world for those of us who have lived through the change, but ultimately a familiar and comfortable world to all human communicators.

29

Work, Information Technology, and Sustainability

Jack M. Nilles

In 1967, as I was riding in a taxi in Munich during my first visit there, I noticed the craters in the sides of many of the buildings – a result of shelling from one side or the other during World War II. It was during rush hour and, just after I had stared for a few moments at a particularly vivid example of mid-forties demolition, the driver turned to me and said: 'You know, this traffic is terrible and getting worse. What we need is another good war! Get rid of all these people!' At the time I thought that he was making some sort of macabre joke, but wonder what he would say now, when the world's population is almost two-thirds greater. Here are some observations on the factors affecting future work and sustainability, as influenced by information technology.

Some problems

As an applied futurist,[1] I spend considerable time analysing current trends of various sorts and trying to devise ways of altering them, generally to point toward more desirable outcomes. This is important in order to put issues in context and to help establish priorities. As a number of books and articles on sustainability have indicated in some detail, current global trends, if unaltered, could lead to some ultimate problems:

- *Ultimate Problem 1:* At some point, the world's population could well become unsustainable, quite likely in some catastrophic way, given a sufficiently long period of continued positive population growth. We can argue at great length about *when* that will be, but *whether* seems certain. We will simply run out of one or more vital resources if we continue growing at anywhere near current rates.

- *Ultimate Problem 2:* When that endpoint will occur depends on a number of factors, such as our rate of destruction of 'renewable' resources, our rate of depletion of non-renewable resources, our ideas as to what constitutes a minimum standard of living for the people in power and for those with the least power, and the economic gap between the powerful and the powerless. The latter, of course, will be the first to go, won't they?

The idea of sustainability is to avoid UP1 either by causing negative population growth – say, by following the recommendations of the Munich taxi driver or by some other means of arriving at a point where population and net resource use are in balance – or by being incredibly inventive and pushing UP2 off to the point where we have discovered a way to ship the excess population to another viable planetary system without using any earth-produced energy. Garret Hardin, in *Living Within Limits*[2], lucidly points out the unfeasibility of massive interplanetary population transportation unless we discover huge new energy resources or teleportation à la *Star Trek* becomes a reality. This leads to:

- *Ultimate Problem 3:* We're probably stuck with the consequences of whatever we are now doing and have done to ourselves and each other over the next few decades.

These Ultimate Problems are mere rephrasings of the laws of thermodynamics, which are often interpreted as:

1. You can't win
2. You can't break even
3. You can't get out of the game

To this I would like to add:

- *Ultimate Problem 4:* No one in a position of significant political or economic power seems to seriously believe that developing workable solutions to UPs 1 through 3 is his/her responsibility.

Unfortunately, we can't just sit there either, if we would like matters to turn out differently. The decision to do nothing is a policy decision. Those who do nothing about altering current trends are implicitly subscribing to the outcomes.

Trends, the courses of events

Now, given those simple laws of cause and effect, let's review a few of the key trends that will influence the existence and nature of work in the future.

Population

The world population is currently at about 5.8 billion and growing at about 1.7% annually. Slightly over 20% of the population is in 'developed' countries and the proportion is falling. The fertility rate is at or below the replacement level in some OECD countries. The 'developing' countries are developing their own technological expertise; even though the experts – or even the literate citizens – in these countries constitute a small fraction of the population, their numbers are significant and are growing at a rate faster than that in the OECD countries – namely about two people in the world of 2030 for every one today. Most of those additional people will be born in Asia, Africa, or Latin America, assuming that current growth rates continue unaltered.[3]

The Work Force

Work traditionally has been categorised into three primary sectors: agriculture, industry, and services. This was a reasonable breakdown for the industrial era, when these sectors were clearly differentiated and of comparable size. However, as the twentieth century wore on, a fourth differentiable sector of the economy arose: the information sector.

Information work is work that is primarily concerned with the creation, manipulation, and / or transfer of information, or work with information machines. In most national statistical compilations it is lumped in with service and, to some extent, included in counts of manufacturing and agricultural employment. Yet, the information sector is fundamentally different from the others in that it contains most of the innovators who are the primary drivers of 'progress.' In the developed countries, such as most of the member countries of the Organization for Economic Cooperation and Development (OECD), the dominant component of the workforce currently is the information sector, generally comprising between half and 60% of each country's work force. The information sector became dominant as a source of employment in the 1950s in the US and possibly slightly later in Europe and Japan. In most of these countries agriculture constitutes less than 5% of employment, with manufacturing / industry at or less than 20% and 'service' filling in the remainder. Roughly half the population in developed countries are in the nominal workforce, so about a quarter of the populations of these countries are information workers.

Third world countries are typically the other way around. Agriculture and extractive industries constitute the dominant source of employment; manufacturing is growing but is generally a small contributor; service in its traditional form (servants, food vendors, ditch diggers, drivers of vehicles, etc.) may be a larger component than information. The population has high levels of illiteracy. The third world also comprises a major segment of the world's population. That is the situation now. That is, the third world workforce distribution is similar to that of the US in the mid-nineteenth century.

But times are a-changing. The developed countries are relatively stable in population and work force distribution. The developing countries have higher population growth and are also making the transition to an information economy, not necessarily by repeating the historical paths of the US and other developed countries. The dynamics of this trend are of vital importance to the developed world and to global economic stability as well.

Aside from the fact that the author and most of the readers of these pages are included in it, the information sector has one other very important characteristic: it is not particularly resource intensive, provided one separates the at-work activities of the information sector from the transportation of information workers. One might say that the historical three sectors represent the fundamentals of sustenance: food, clothing, shelter, and physical tools. The information sector represents much of what happens after the fundamentals are taken care of, although even the fundamentals have information components.

Technology

One of the main influencing agents in the dynamics of the work force is information technology: computers and telecommunications. Both of these sets of technology are growing in performance per unit of price at annual rates between 25% and 30%.[4] This means that, for a given level of investment, computer power grows – and the cost of communicating decreases – by an order of magnitude about every seven years. This has been the primary engine for the appearance of personal computers on or near most desks in the developed world, as well as the recent, seemingly spectacular, growth of the Internet.

Powerful computers are becoming inexpensive and the cost (if not the price) of telecommunications is becoming trivial. Of course, 'trivial' in a developed country does not have the same meaning as it does in a developing country. But the price of telecommunications has been set in most countries by governments rather than the marketplace. As telecommunications utilities become largely privatised over the next few years, and as competition increases in that market, look for prices that more closely track costs of delivery. One of the more powerful factors in this regard is the growing ability of wireless communications (via microwave relay or satellites) to deliver sophisticated telecommunications to sparsely populated areas that are practicably unreachable with wire-based technology.

Many developing countries have understood these relationships and are busy developing both the telecommunications infrastructure – both wired and wireless – and the technology distribution systems to avail themselves of their potential economic benefits. Before 2010 there will be more telephone lines in the rest of the world than in the OECD countries. The same is likely to be true of Internet host computers.

Because of these technology trends, the world is indeed becoming a global village in the sense that *it will soon be feasible for anyone to have some form of instant communication with anyone else, anywhere else in the world, at a price that is affordable for a substantial[5] fraction of the world's population.*

Economics

In most economies, the societal link between work and survival (at whatever level) is firm and largely unquestioned. That is: the absence of a worker in the household means no survival for the household. In the developed countries, information work is the predominant form. As shown above, physical sustenance and mobility – food, shelter, clothing, transportation – are provided by less than 25% of the work force and that fraction is shrinking. In most developing countries, agriculture is still the dominant source of jobs / tasks but – as technology diffuses – the trend is definitely toward some version of repetition of the rise and subsequent fall of the industrial revolution.

All of these economies are based on the repetitive process of production of goods or services and distribution to consumers thereof via some form of marketplace and medium of exchange. In traditional, pre-information economies this process usually involves the physical transfer of something bulky from producer to ultimate consumer. That is, some form of energy consumptive, usually polluting, transportation system is an integral part of the process, at least since the end of the age of sail. In a fully developed information economy, the primary form of 'transport' is telecommunications, the raw materials (data) and the products (information) – but neither the producers nor consumers are transported and the transportation media are relatively pollution-free and low in energy consumption.

Infotechonomy

In an information economy, information technology provides for a growing amount of *location independence* for the tasks to be performed by a growing proportion of job holders. This location independence can be local (telecommuting), regional, or global (teleworking at various scales). Telecommuting is well documented, I have written a few hundred thousand words on it myself. Less well documented is broader scale teleworking, particularly of the trans-border variety, but it is clearly increasing. Further, there is a gap between the skills needed to be an expert user of infotech and the training / abilities of prospective users. To some extent, the gap is narrowing as the machines (and their software) get 'smarter' and / or their users learn the proper incantations. Yet, the gap may be widening at the upper end. The information elite vs. the information deprived will be a continuing issue globally. The information elite will definitely (if they choose) be in the well-paid-for work part of the economy.

One way of summarising this is: Basic physical survival, can be taken care of by either a small proportion of workers or a small proportion of the time of a larger set of workers. Physical services, such as gardening, flipping hamburgers at MacDonald's, etc., also employs a diminishing fraction of the population as more of even these activities are automated or priced out of existence. So what are the rest of us to do? Take in each other's intellectual washing, as Joseph Weizenbaum put it? The design of non-survival forms of work and the acceptability of paid-for non-work become the dominant issues.

I have often said that it is fairly easy to forecast the demise of certain job types, such as (traditional) secretaries, clerical workers, typewriter manu-facturers, buggy whip braiders, etc. – i.e., structural unemployment – but difficult to forecast and describe the jobs that have yet to be invented – structural employment. However, it is clear to me that a major source of future employment is a direct consequence of a primary human attribute – curiosity – and information technology. The more information we get, the more we want to have. The more data we get, the greater our difficulty in sorting it out and converting it to information. So far we seem to have managed to increase complexity of our work faster than we have man-aged to automate it. Still, we either have to invent those new jobs or make major changes in how our economies work.

The crucial issue is whether the introduction of a new technology merely eliminates jobs – the negative sum outcome – or whether it creates condi-tions that will ultimately create new opportunities and new jobs – the positive sum outcome that is required if the traditional view of work is to remain as the dominant goal. Typically, a new technology is used solely to replace some process that is already established in the organisation and the immediate result is to displace workers or reduce demand for more workers. Only later, usually, are new uses invented for the technol-ogy and new demand created for improved versions, and so on. In a mature industry, that is, one in which the market is fixed in size, the likely long term consequence of the technology is structural unemployment. Only if the technology results in new markets will employment increase. Hence, if information technology is to result in an increase in traditional forms of work, it must operate more in the positive sum sector than on the negative side of the books.

One of the possible drivers of positive sum changes will be what I term evanescent and dispersed or distributed organisations; organisations that form and flow, connecting and collecting expertise in response to changes in market/environment conditions. Many of the project teams that have resulted from initiatives in the European Commission typify that form of organisation, including three with which I have been involved. The mem-ber units, even the leadership of the group, can change as conditions warrant. There may be some constant components, to help insure stability and some form of group identity, but many will be entirely ad hoc. Large multinational companies have been doing this for some time (Benneton as an example) but even small organisations can play. In fact, there are obvious advantages to global evanescence, particularly in areas where rapid response is crucial; keep shifting the tasks east (or west) as the clock moves on. The positive sum aspect of these organisations is that they enable new work combinations and opportunities to exist that would not otherwise be possible, absent the technology.

The good news is that these organisations exist already. The bad news is that they mostly comprise the information adept. The benefits have yet to accrue to the less skilled and less entrepreneurial, who will increasingly be the victims of structural unemployment – unless the economic rules change.

These examples also illustrate another key point: location independence also means border independence. A primary potential impediment to the establishment of these organisational forms at the supraregional scale is the lack of uniform technology standards and regulatory / tariff barriers to trans-border data flow. The putative advantage of these barriers may be to protect against the lower wage demands of the experts in developing countries. The disadvantage is that they impede information flow within the barrier countries and increase the tensions between the haves and have-nots.

Energy and Resource Economics

All of the forecast graphs previously shown were made under the assumption that there are no physical limits to growth in the period considered. This isn't necessarily valid. Already we are seeing the outlines of limits in desertification, pollution, and global warming. We are also seeing reports of the increased extent to which oil exploration companies must go to provide us with cheap fossil fuels. Aside from the issue of finding suitable energy resources,[6] there is the problem that most of this energy eventually winds up as heat, exacerbating global warming trends.

Although the rate of depletion of the world's mineral resources (excluding fossil fuels) is diminishing, thanks to widespread recycling, food resources are not keeping up. For many years problems of starvation in third world countries could be attributed to transportation failures; the food existed but was in the wrong places. We are already seeing signs of absolute deficits in food resources. For example, in 1995 the Peoples Republic of China became a net importer of grain for the first time. Mass starvation, economic sanctions, ethnic cleansing, and other such inven-

tions are tried and true, if not desirable, ways of stabilizing the population. AIDS or some similar plague, though less tried, may be just as true.
Finally, starving or otherwise stricken people are more likely to support
acts of terrorism that also reduce the population, although not necessarily near the homes of the terrorists. In short, population growth and
resource depletion are on a collision course.

The Consequences

This growing info-tsunami has important economic and environmental
consequences for all of the world's countries. For developing countries it
means that they may be able to sell the products of their skilled citizens to
buyers elsewhere, while keeping the income at home, thereby increasing
their rate of development and economic growth. Of course, this is already
happening in many developing countries. The point is that the scale of
such telework transactions may materially increase in just a few years.

Similarly, developed countries, with their large pools of expertise, may be
able to successfully expand their knowledge sales on a broader scale via
telework to the developing countries. This, too is already happening as
developed countries export their skills via telework to Pacific Basin countries, for example.

Yet the scenario is clearly possible that the developing countries, with
their growing populations, may be able to 'out-expertise' the developed
countries in a decade or two, causing major shifts in the distribution of
both physical and intellectual capital. A great variety of outcomes of these
trends is possible. They depend on actions taken by governments and
private sector institutions regarding education and training, regulation,
taxation, investment in infrastructure, and trade policies, to name a few
areas of interaction. Whatever the outcome, telework will play an increasingly important role. It is important to remember that, by 2010, there may
be as many information workers outside the OECD countries as there will
be within them. In the intervening years there will be significant changes
in global trade patterns and the distribution of material wealth as a result
both of the population shifts and technology ubiquity.

Telework's environmental impacts are also important. Computers and
telecommunications technologies are relatively small consumers of both
mineral resources and energy, compared to traditional industrial products. Teleworkers do not use transportation resources as much as other
workers. Hence, per capita energy consumption is reduced. Thus, the
continued worldwide transition to an information economy may at least
reduce the rate at which we approach those ultimate problems, giving us
more time to find solutions.

However, it is still the case that no one is in charge. These telework actions will take place largely as a result of individual decisions, although government policies and actions can influence them to some extent. Emphasis on, or downgrading of, education is one of the key policy areas. The economics of telework, although not discussed here, have strong appeal to ventures that are based on the profit motive and the existence of free markets. What is in question is not the nature of the outcomes, but the rates at which they will occur.

Still unanswered is the question of whether enough new jobs will be invented to purposefully occupy half the world's population in coming years. Quite conceivably a new paradigm of the relationships between work and rewards is needed, coupled with a deeper understanding of the laws of thermodynamics and the constraints placed upon us by spaceship earth – not to mention a little genetic reengineering. Further, it is no longer wise to consider these issues on just a country-by-country stage.

It is way past time to get organized and find/develop leadership in the search for both better understanding and practical solutions.

References

1. That is, I try not only to see where we're heading, and why, but to discover what to change, how much, and when, in order to arrive at a better outcome. My research on telework began in 1970 as an investigation into the problem of urban traffic congestion.
2. Oxford University Press, 1993, chapter 2.
3. This is an unlikely assumption, since growth rates in the developing world will decrease for one reason or another. However, *when* that will occur is uncertain.
4. 'Moore's Law,' which describes these empirical facts, will start to flatten shortly after the turn of the millennium as microelectronics comes up against fundamental quantum limits to growth. There are lower, as well as upper, limits to growth.
5. I'm not yet sure how substantial 'substantial' is. We are just beginning that study.
6. An interesting example of path dependence, found in an article in *The New York Times Magazine* ('Why the Best Doesn't Always Win', by Peter Passell, May 5, 1996) is the dominance of the light water nuclear reactor over a potentially safer and less polluting gas-graphite system. The light water reactor was the most compact and therefore most suitable for use in US Navy vessels.

30

How Will Future Information Technology Affect Me? – A Personal User Perspective

Karl M. Wiig

Introduction

The larger part of earth's population has been touched by information technology (IT) in one way or another for quite some time. The rapid development of IT capabilities has led to mushrooming IT-based and IT-related services in most aspects of modern society. Developments are continuing with no end in sight, in many areas, and at an increasing pace. More than four billion people already are significantly affected in their daily life, the way they work, the way they shop, they way their societies provide services and are operated, and in many other ways. The influence is greatest in the developed nations, but is also noticeable in many less-developed areas.

In general, IT focuses on devices and functions that manipulate, transmit, store, access, distribute, and present information. From a functional perspective, IT is expected to support a wide range of activities and purposes in personal lives, business and industry, and society. We can already glimpse many near-future changes that IT might bring about. Predictions of longer-term changes are pure speculation since they are likely to result from innovative individuals' ingenious inventions and therefore tend to break with prior thinking and may not be foreseen.

This chapter explores how future IT applications might affect us, as users, in various areas of our lives. It will not speculate on the precise nature of new technological solutions and the theoretical breakthroughs that may take place. However, implicitly, such assumptions are the basis for depicting what might happen. In the following we present a few examples of how people may use IT as perceived from this author's perspective. Inevitably, that results in a biased, limited, and often erroneous view.

Changes in Our Personal Life

The influence of IT in our personal life has barely scratched the surface compared to what is likely to come. Many functions will be created by popular demand as the public, business, and government discover how useful and valuable IT functions can be to support daily life and to promote functionality of institutions. However, as we already have experienced in many areas, creators of IT products are prone to bloat their products with many advanced features in a technology push mode. This tendency will continue to make many IT functions complex beyond the average user's capability, and may make many products unusable by the general public, hence creating societal divisions between IT-aficionados and common folk.

In the home, there will be IT-based functions for many purposes. The central IT utility is the FIMS (Family Information Management System) which keeps general records and supervises and coordinates other information functions. These functions are integrated to provide seamless support and make daily life effective and simple. FIMS support several functions throughout the house, ranging from controlling air flow and temperature to maximize comfort and minimize energy consumption, to managing the security system.

Each family member has two IT devices the PIA and PWT, and a personal information environment, the PIM:

- PIA (Personal Information Assistant) is a wearable computer the size of a large watch and worn as a bracelet by each person. It is ruggedized, self powered, has considerable nonvolatile memory, input sensitive display, universal networking capabilities, and recognizes its owner who can be its only user. It also communicates with mono and stereo heads-up display and eye-sensing pointing devices that support VR (virtual reality). Specialized PIAs are worn by people with certain health conditions to monitor vital signs and other variables. The PIA also acts as a personal Internet-based cellular telephone and videophone and can receive and display video messages and TV programs but only 'on the fly' since it has limited video information storage.

- PWT (Personal Work Tool) is a portable computer – which ultimately may be incorporated in the PIA – that people use at home and at work and school. It is lightweight and is each person's information station for all purposes in any setting. PWTs come with varying capabilities, all are ruggedized, have large memories, highly readable flat and heads-up displays (a full HDTV window occupies half of the display), key, voice, other types of inputs, and universal networking capabilities. PWTs also support VR and can only be used by their owners. The PWT is used for all types of information-related work. It is used to capture drawing and writing, directly or through dictation. It is used to explore information resources and work with educational materials. It is used to view TV, movies, and other video programmes, and to read magazines, newspapers, and books.

- PIM (Personal Information Manager). In addition, each family member has a PIM which physically may be part of FIMS. It serves as each person's master information server, acts as message center, and provides backup for PWT and PIA. When needed, PIM communicates with PIA and PWT over networks.

To illustrate examples of how IT may affect us, we follow a typical family – the Shaws – in part of their daily lives, at home, at school, and at work.

The Shaws

The year is 2015. The Shaws are a typical family and are as affected by IT as everybody else in North America, Europe, and other developed nations. John (39) is a civil engineer for the city, Mary (38) teaches junior high school history, Anna (14) is in ninth grade, and Billy (8) is in third grade. Their house is twenty years old, was recently retrofitted with fiber, infrared, and wireless communications, and connected to the local information network as have become the standards everywhere.

Kitchen and Household Management

The HMS (Household Management System) is a specialized function within FIMS. Its kitchen function is networked to all kitchen implements and it controls appliances all around the house.

John is planning the family meals for the coming week. The HMS has a fair record of food already in the house by recording arrival of food, sensing food storage bin levels, and deducting what is used for meals. Barcoded SKUs on packages are read when adding or removing food items but the records are not accurate. (Billy and Anna often remove items without recording!) The HMS proposes menus from its recipe file based on family preferences, what was served recently, and what is in the house. FIMS has informed the HMS about special family health requirements that affect the diet. By special requests, John makes changes and generates the shopping list for needed items after the HMS reminds him to check what has been 'pilfered' and sends an email to the store for home delivery.

The Shaw's kitchen has computer controlled dispensing and level sensing of dry consumables (e.g. cereals), and liquids (e.g. refrigerated milk). Many ingredients are dispensed directly into the appliances. Ovens and other food processors read labels on food packages and have embedded information on how to prepare the content.

The HMS tracks the use of general consumables and adds needed dishwasher soap, paper towels, or other items to the list as required. It also tracks water and HVAC filter conditions and adds new filters to the shopping list when needed. HMS sends shopping lists to the different stores that carry the items along with payment and delivery information. The stores later deliver the merchandise to the door when someone is at home.

Start of the Day

In the morning, everyone grabs their PWT that has downloaded personalised information during the night. While having breakfast, Mary and John review messages and read news items provided according to their preference profile. They also browse the general news – with advertisements – to catch anything of interest outside their profile.

Anna and Billy watch TV programs on their PWTs and check messages from school and friends and after checking with his mother, Billy accepts by email an invitation from his friend to play after school.

Towards the End of the Day

For dinner, the HMS alerts whoever is home when it is time to transfer food items from storage and indicates how they should be prepared. The HMS programs the oven (or other food processors) to prepare the selected recipe. Each device is activated by today's cook after adding manually handled ingredients. Some meals are prepared without human intervention, particularly on days when nobody will be home early. During dinner the discussion focuses on where to go for the upcoming long weekend. Anna uses her PIA to explore which spaces are available at their favorite camping place.

After dinner, all go to their PWTs. Billy solves math problems with a CBE (Computer-Based Education) application. Anna downloads video material that she edits and incorporates in a history report she is completing. Mary reviews and grades student projects she downloads, and John dictates memoranda for his job. Later they all sit down together and play a version of 'Battleships' by networking their PWTs. While playing, John watches a baseball game and Anna and Mary are watching and discussing the same video program on their PWTs. Billy focuses on winning.

Learning May Be Easier – It Will Certainly Be Different!

SIE (School Information Environment) provides support for all administrative and educational functions within the school. It is networked to the Internet and to the school system's intranet server which also houses school and student records, monitors student progress, and has educational material for all curriculum topics for all grade levels.

Students and teachers bring their PWT (Personal Work Tool) to school for use as their work stations. The PWTs are networked within the classroom and collaborating group and to the SIE. Students are not issued paper-based school books. Instead, they use downloaded educational materials that they work with on their PWTs. Since students at any time may team with others anywhere on the campus, SIE also tracks where each student is by sensing where their PIAs and PWTs are connected to the network.

Given the power and flexibility of content-focused IT, the nature of education has changed. The objective of education in all grade levels has been shifted from rote learning and memorizing. Instead, the objective is for students to build deep understanding – script and schema mental models – of the subject matter. Where tacit, automatic capabilities are required, such as for performing arithmetic operations, deep understanding is linked to automatic procedural mental models through strong associations. For example, to achieve automaticity in basic mathematics, interactive computer-based applications are used to lead students repeatedly through 'word problems' that require interpretation and formulation before the student can perform the calculations.

At school Billy is chronologically in third grade and Anna is in ninth grade of the local public school. Although years apart, they still are taught in similar ways and use their PWTs extensively. In all their subjects, literature, geography, physics, history, and so on, they collaborate with other students, often from different grades, and are given subject-related projects which they might choose because of interest and are encouraged to explore and research on their own. To find information, students use Internet and many proprietary databases that their school has acquired. Their teachers act as process mediators with substantial content knowledge but are not required to be the content authorities of yesteryear. Instead they act as guides and mentors to help students build understanding of the subject matter and to apply critical thinking.

The IT-based learning environment has made it possible for Billy to advance rapidly in math and physics where he has special aptitudes. In history and literature he lags behind the class and Mrs. Lorentz, his classroom teacher who monitors the progress of each student, provides him with special learning materials. She has also teamed him with Martha, a fifth grader with a knack for helping. Billy himself helps Mathew and Tom who are in the same classroom with math. They network their PWTs and using a collaborative environment, they work on the problems with Billy explaining in writing and by discussing until Mathew and Tom feel they understand.

Billy, Anna, and their schoolmates use CBE (Computer-Based Education) applications in many subjects. CBE incorporates technology to match the educational path with the cognitive style of each student, detect student misconceptions that can be corrected dynamically, and provide opportunities for students to provide feedback or pursue sidelines of interest. The CBEs help students develop good concept understanding, sound mental models, and where appropriate, automated skills such as in basic arithmetic, reading, and so on. The CBE applications are downloaded from the SIE to the PWT for use by the student at school or at home.

Last month, Anna and Billy went on a two week trip with their parents in the middle of the school semester. Their teachers gave them assignments and downloaded CBEs for them to work on their PWTs while they were away.

At work, John and Mary also receive sophisticated CBE applications that they work with to keep abreast of changes and to upgrade themselves for new challenges.

In 2015 IT plays a major role in education with miniaturised technology, AI, virtual reality, and other sophisticated capabilities. The reliance on these capabilities could easily be carried to extremes – to the point where students would lose touch with the real world. The educational practices are designed to avoid that and provide situations and laboratories for experiencing social, biological and physical phenomena in reality.

Health and Professional Services

As in the 1990s, professional service providers, hospitals, physicians, dentists, lawyers, and so on have office computers but with more extensive capabilities. Beyond traditional client record keeping, billing, and so on, the functionalities include expert knowledge bases, AI reasoning capabilities, and networking with expert sources to deal with unusual challenges. Craft providers like plumbers, electricians, and so on also use similar office computers to deliver highly effective and competent services.

Anna has hay fever and visits her allergist, Tom. During the last weeks, Anna has recorded information on her PIE bracelet about her symptoms using an intelligent guide specific to her allergy. Her PIE contains this information and her full medical history. The new information is transferred automatically into the MOC (medical office computer) which evaluates and highlights aspects relevant to today's visit. The MOC also obtains the latest medical information from the National Institutes of Health (NIH) to explore opportunities for improving Anna's treatment.

When Tom sees Anna, he reads the MOC summary, examines her, has time to talk with her, and obtains information on her present situation and complaints. He is quickly able to evaluate her situation and decide on the most effective treatment by combining his understanding with inputs from NIH without missing anything that might be pertinent. Tom dictates the present diagnosis and treatment to MOC which incorporates it into her record. Before Anna leaves, the new information is downloaded into her PIE bracelet. MOC contacts her FIMS with the information which also specifies medication and diet suggestions. MOC contacts the pharmacy with prescription and the insurance company to settle billing. MOC and the pharmacy computer obtain other information from FIMS to avert medication conflicts and make recommendations, when required, about dietary supplements and over-the-counter medications.

On the home front, FIMS has daily contacts with the pollution information authority to identify when Anna needs medication. FIMS also has information on all the other family members' health conditions and integrates that information when determining all its other actions.

When Anna, or the other family members visits their family physician, Ellen, for regular checkups, the routine is similar. The PIE is downloaded into the MOC which contacts the NIH system for the latest relevant information and Ellen can quickly obtain good understanding of the situation while also taking into account the inputs from other healthcare providers that have been seen.

Money Management

The Family Money Manager (FMM) resides within FIMS and uses embedded financial, investment, and tax planning expertise and AI-based reasoning capabilities to assist the family manage its finances. It networks with financial institutions with whom the family has business and with information service providers in general, over the Internet, and other channels.

The Shaws use little cash and no cheques. John or Mary use FMM to transfer budgeted allowances and other spending money from their accounts into Anna and Billy's PIEs for use when at school or shopping. John and Mary also maintain up-to-date information on their credit and debit accounts in their PIEs for use as they need it.

John and Mary use FMM to assist in managing their financial resources. FMM receives information on all the Shaw's transactions. From this information it generates categorized reports and provides recommendations for handling the finances differently. FMM communicates regularly with their financial institutions (banks, insurance companies, brokerages, etc.) to obtain the current state of accounts and execute authorized transactions. They use

FMM to search for the best insurance policies and explore opportunities for educational and retirement investments. FMM employs intelligent agents (IAs) to seek out investment options whenever needed and constantly monitors the performance of the Shaw's investment portfolios and either alerts them to initiate action or makes pre-authorised changes. Sometimes, such as when they plan to buy a new car, they use FMM to look for available financing options and alert them when interesting situations are located.

Anna and Billy also use FMM to manage their savings accounts.

IT Will Influence the Way I Work!

In the year 2015, each company and organisation is supported by a CIE (Central Information Environment) which serves many purposes. It provides traditional information management services for the organisation. It maintains networking between the organisation, its employees, and customers and suppliers. It is the general information repository. It is also the repository for lessons learned and structured intellectual capital assets ranging from educational materials to reasoning behind systems and procedures.

In most organisations knowledge workers operate in the same manner. They often telecommute two or three days each week, but when required are instantly available through their PIEs. However, personal interactions are recognised to be important and employees regularly need to meet face-to-face. Normally, they do not have fixed offices but when at the office use rowing desks or conference rooms where their PWTs network with co-workers and with the CIE. All normal work materials are resident in their PWTs. Knowledge workers who serve clients directly, customer service representatives, for example, normally have fixed workplaces. However, some of them also telecommute.

Communications are mostly electronic. Occasional paper-based communications are scanned and distributed electronically to those who need them or stored for later access.

The Work Place

Each day, John receives many communications of varying importance. He lets his PWT organise them according to topic and priority. Some items are discarded immediately, general items 'for your information' are filed for further reference, while items considered important are summarized using natural language understanding and brought to his attention.

At this time, John's major responsibility is the design and construction-planning for a new highway interchange in the city. He is present in the office for personal meetings but has no fixed office space. His work takes him to several construction sites and he may work at home two days every week. A couple of days a week he goes to the office to 'be around' his associates to learn what they do and think by meeting them and have informal interactions.

John collaborates on the design of the highway interchange with vendors, community representatives, and coworkers in many different locations. He has had preliminary personal meetings with the city's Parks Department and other groups to discuss difficult issues. Today he meets with eleven representatives – six are present, five are remote – to hopefully resolve several of the issues. The meeting is held in an open meeting room equipped with networking for each participant's PWT and with the CIE. The CIE provides all relevant background information as required. More importantly, it provides the meeting protocol and collaborative work environment (development of the 'groupware' technology from a decade earlier). Based on the preliminary discussions, several of the participants have created brief multimedia reports to summarise their perspectives on the issues and these reports were distributed and reviewed by all participants and many of their associates.

As a result of the independent workstyle, John is accountable for what he achieves – not for how many hours he spends on the job. Hence, his position is different from many knowledge workers who work in industries such as manufacturing and chemicals where they need to be present for specific periods.

In his job, John has good understanding of what is expected from him. Whenever city managers make decisions that result in strategic or tactical changes, the CIE downloads the changes to his PWT. It notifies him and sets up links to related items to explore implications further. He also has available a clear 'service paradigm' that outlines his agreement with city management on how they expect him to conduct himself in his position.

The Quality of Work

Advanced IT will lead to intelligent automation of some less complex work as indicated in Figure 1. Also, the effort to deliver present day service paradigms will be reduced. However, work can be expected to change – to become more demanding – to satisfy increasing market requirements for new product features and increased capabilities of services. Although sophisticated IT will improve job-related understanding through better script and schema knowledge, work will expand to take advantage of the new capabilities. Advanced IT will provide better communication and higher quality information allowing knowledge workers to accept increased responsibilities, by feeling more confident and having better understanding of work to be done. With the better knowledge support provided by new generation IT, more jobs will be done right the first time, adding to confidence and job satisfaction on the inside, and better market acceptance on the outside.

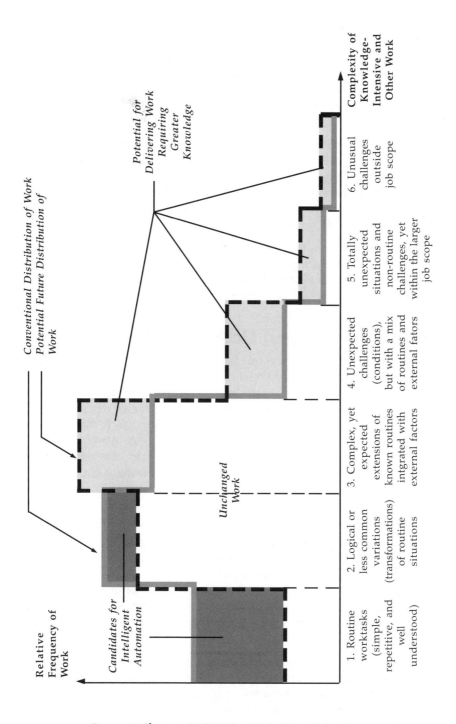

Figure 1. Changes Will Make Work More Complex.

Going Shopping

Electronic shopping is pervasive and provides opportunities to consider more possibilities. The new ways of shopping make possible better fits to requirements and with more sophisticated checking and monitoring, it is also more secure than earlier. Improved worldwide communication has changed personal shopping to transcend geographical limits.

The Shaws use many shopping channels. Whereas they normally let their HMS buy food and household supplies, they also go to the supermarket to 'squeeze the oranges'. Mary and Anna prefer to shop for clothes at local stores, although they also buy many items over the Internet from merchants anywhere in the world. Catalogue sales are exclusively electronic and purchasing over the Net is greatly facilitated by 3-D and VR.

When buying clothes, their measurements are scanned to 'build' garments that fit perfectly and are guaranteed for quality and workmanship. For large items or professional services, they obtain customer references and communicate over email to check out vendors.

Public Services and Decision Making

Advanced IT introduces several changes in public life. Information web sites provide targeted and current information on public programs, social services, numerous specific issues such as rights of individuals under different conditions, and so on. Public agencies make detailed reports available to all citizens. Non-restricted databases are made publicly available. Extensive navigation aids using intelligent query agents and 3-D data visualisation provide assistance and are backed up by commonly understood ontologies. As a result, people are better able to make good personal decisions by having access to high quality information on public services and issues that affect them directly. People are also able to participate in public decision making with greater understanding and impacts.

John surveys perspectives of all residents in the city about the new road exchange he is responsible for and other traffic-related issues. The survey is part of a monthly opinion gathering and is communicated to everyone over the city's intranet. The results are tabulated and analysed automatically and made available to all and they are also considered when making design decisions. Rate of response to these surveys is always high and provides direct inputs to city decision making.

John and Mary frequently access the city's and state's databases to explore how changes might affect them and what they should do. For example, they submit different scenarios to the tax advisory service to obtain suggestions for how to arrange their affairs differently. They also follow community propositions and provide feedback on their preferences to influence final decisions.

Anna is interested in finding new and interesting activities outside the home. She does not quite know what she wishes to do, but searches the Net to learn more about groups that she might wish to join.

What May It All Mean?

IT devices will undergo extensive developments during the next decades. However, to make the Shaw's world a reality will require enormous efforts to create operational and content software. The software needed to provide the reliable, seamless functionalities required must be of a quality and comprehensiveness rarely seen in the 20th Century. However, that will only secure the operational functionality of IT. Even greater efforts will be required to embed content – the knowledge and expertise, procedural methods, information, and facts – into the many systems that are needed to create and support such an environment.

The IT scenarios indicated above will lead to significant traffic over all channels. Internet traffic will be large compared with today's levels to support the automatic monitoring and exploratory functions that users are likely to initiate and the amounts of video and other complex information that will be transmitted.

The broad use of CBE (Computer-Based Education) will make it economical to embed considerable sophistication into these applications. That makes it possible to deliver more comprehensive, deeper, and easier to understand knowledge to students, young and old, than conventional teacher-conducted education can provide. Hence, students can be provided with greater understanding in wider domains while still being capable to handle mechanical tasks such as the '3Rs' of Reading, wRiting, and aRithmetic.

Beyond these considerations are the requirements to create economic, social, and political environments to develop and support the seamless integration of how consumers, service providers, and institutions function and work together. Intellectual capital must be traded, even shared. Economic mechanisms must be devised to facilitate efficient and equitable exchanges. Security measures must be established to protect privacy and to minimize the potential for fraud and system failures.

From the users perspective, the advanced IT services of the future will provide many advantages. With little effort, IT provides the Shaws with a large span of reach for learning, following current events, arranging their financial affairs – including entertainment and shopping. In total, the hope is that IT will provide a low friction worldwide environment that will make daily life less cumbersome and more fulfilling at all levels of society.

31

The Future of Scholarly Skywriting

Stevan Harnad

My own corner of what I've called the 'PostGutenberg Galaxy' is a relatively tiny one. Alongside the video, audio, commerce, adverts, chat and erotica, the scholarly/scientific portion of the Net is like the flea on the tail of the dog. But that flea is destined for great things, and humankind will be the beneficiary.

First it has to be clearly understood that the flea differs from the rest of the dog in one crucial respect, but this difference means that its future in cyberspace will differ from most of the rest of its inhabitants.

Let's drop the flea metaphor. The flea is the refereed journal literature: At least 14,000 periodicals are dedicated to publishing the ideas and findings of researchers in all the scholarly fields – the sciences, arts and humanities. Now the critical difference in question is that the authors of all those articles – unlike the authors of books, and of articles in trade magazines – do not write their papers for royalties or fees; they give them to their publishers for free. The only reward they seek is the eyes and minds of their fellow researchers, so that their ideas and findings can have their full potential impact on the future course of research.

In the Gutenberg era the only way these special authors could bring their ideas and findings to the attention of their peers – present and future, the world over – was by treating them exactly the way trade authors did: They gave them to a publisher, who paid the considerable costs of printing them on paper, and then recovered those costs, plus a fair profit, by charging for access to the 'product' (even though these non-trade authors did not seek or get any of the proceeds from the receipts).

Despite the expense and inefficiency of disseminating their work through print on paper, and despite the deterrent effect of restricting access to only those who could and would pay for it (usually in the form of institutional journal Subscriptions, but lately also through shared arrangements such as institutional Licenses or Pay-per-view in various forms – let us call these S/L/P), scientists and scholars were well-served by this system in the Gutenberg Era, mainly because it did disseminate their work, and there were no alternatives.

The PostGutenberg Galaxy offers an alternative. Just as in the paper era these special authors had (at their own expense) sent free reprints of their papers to everyone who requested them, so today, these authors can self-archive their papers publicly on the web, so all interested heads and minds can access them from anywhere for free.

In 1999 this is already being done by 100,000 physicist-authors and 35,000 daily physicist-readers in the Los Alamos Eprint Archive, created by Paul Ginsparg in 1991. It will be an interesting matter for historians to unravel why, having been led to the water, researchers in all the other disciplines have been taking so long to get around to drinking, but it is a foregone conclusion that they will catch on and do so, sooner or later.

Once they do, the entire refereed literature will be available to every researcher everywhere at any time for free, and forever. No more trips to libraries to chase down a reference (assuming your library can afford to subscribe to the journal in question at all), no more distinctions between academic haves and have-nots when it comes to being able to keep abreast of the journal literature. In fact, the literature will be seamlessly interconnected, with hyperlinks from every paper to every other paper it cites. Still more important, the probability that each paper reaches its full potential readership (and hence makes its full potential contribution to the future course of knowledge) will be much higher. Today, the average journal article is cited by no one and read by few. All papers will not become best-sellers as a result of public self-archiving, but they will certainly have a much better chance of having the full impact they were destined to have once the access barriers of both paper and its costs are out of the picture.

Removing access barriers and putting the entire literature at everyone's fingertips, however, is just the first step in the scholarly revolution that is afoot, although it is the crucial one (waiting only for the academic thoroughbreds to take to the water). For once the barriers are gone it is not only scholarly access that will skyrocket as it never could before, but so will scholarly interaction, the creative interplay between those idea/findings and those eyes/minds that this whole special subfield of publication has always been about. For here the handicap was not just the cost barrier, but also the time barrier.

Human thought evolved hand in hand with human language. The speed of thought and the speed of speech are of the same order of magnitude (if not even more closely coupled). This is because speech evolved in the service of interactive thought: Interdigitating ideas were better than solipsistic ones, simmering in just one cranium. Science and scholarship evolved out of the oral tradition of exchanging ideas and findings by word of mouth. Writing created a permanent record, which increased both the scope and the reliability of the tradition. The written word not only has a broader reach, in time and space, than the spoken one, but it is also incomparably more disciplined, answerable, and hence objective.

Never mind; we are not here to sing the praises of prior revolutions (speech, writing, print), but of a future one: skywriting. For every self-archived paper on the web is like a piece of skywriting, visible to one and all, today and forever more. Still more important, skywriting is there to have further skywriting appended to it (rather like serial graffiti on a public wall, although the analogy is otherwise unflattering, and irrelevant when it comes to *scholarly* skywriting, as opposed to mere Netnews-style chat groups that are really just global graffiti-boards for trivial pursuit).

The paper journal literature could never support interactive skywriting; its turnaround times were simply too slow. By the time a published response to your work appeared, you could no longer even remember what it had all been about! This is nothing like a live interactive conversation – and a good thing too, because live conversations are rambling and unconstrained, alright for conferences and symposia, but not what one wants a permanent archive to consist of (as anyone who has read transcripts of live interactions will agree).

But skywriting offers a hybrid possibility, not quite like anything that came before it: much closer to the live interactive tempo of spontaneous on-line speech (and hence online thought), yet retaining all the virtues of the written medium (formality, discipline, objectivity, publicity, corrigibility, permanence). For not only is it possible (within minutes, if one wishes) to post a skywritten comment in response to a piece of skywriting – something completely impossible in the paper medium (or even in the online medium as long as it is criss-crossed with access barriers from S/L/P), but it is also possible in the online medium to make a piece of skywriting come 'alive,' even if its author is deceased, and to interact with it using all the online dialogic resources for which our brains are specially adapted.

I am referring to a feature that we have all become accustomed to using in the past two decades of email and mailing lists without realising just how revolutionary it would be if it were being used formally by researchers at the highest level of peer interaction, rather than just in private messages to friends or in public chat groups in which the blind lead the blind. The quote/commenting capability with online digital texts – the convention we use in email, in which excerpts from your message to me appear indented and preceded by a '>' sign, followed by my comments (and the possibilities of further iterations and embeddings of this, for which we have not yet developed the codes or the modes, but will) – is the hybrid capability I have in mind, but its revolutionary potential will only become apparent once the peers of the realm have finally opted to drink from the water to whose brink they have now been led. My guess is that their productivity will increase by an order of magnitude once they deign to drink.

32

From the Satisfaction of Basic Needs to Information Literate Societies

Ina Fourie

1. Introduction

In the new millennium information-related aspects will offer new opportunities, new challenges and new problems to deal with. Timely
identification of these can help us to gain insight, and to prepare ourselves and society to survive and prosper in the new era. The complex
nature of the future of information should not be underestimated: technology, economics, culture, religion, education, psychological behaviour
and human reactions will all have an impact. We also do not know which
resources will still be available (e.g. wood for paper, electricity supplies,
steel and optical fibre) or where people may find themselves – some of us
may be on other planets!

The future of information seems even more complex if we consider the
contradictory nature of our society. This society can be viewed on a continuum ranging from societies based on oral literacy and illiteracy to
highly technological societies, from very poor societies to the very wealthy.
In these societies, individuals will also find themselves on similar con-
tinuums. Furthermore, societies and individuals will differ in their needs
for information, their expectations of technology and information systems and their skills in accessing, interpreting, analysing and applying
information. People will also continue to differ in their information search
behaviour and their attitude to information. Is information considered a
valuable resource? Is information an essential resource in keeping a competitive edge? Does it have prestige value? Or is it merely the answers to
questions about safety, protection, housing and health care?

No common or general future can be predicted for information. The situation in which we find ourselves is much too complex. Each sector of
society will be affected differently, and because of the effect of information
will react in different ways to information and use it differently. There are,
however, detectable trends. The way we view the information needs of

different sectors of society will influence how we act to address the underlying issues, which will in turn influence how we shape the future. Our best option, therefore, for dealing with the future is to use society's needs to set aims for the future, and to link these to identifiable technological trends.

Technological developments will have a definite impact on the future of information. Experts seem to be fairly certain of the direction some of these will take in the near future, and of the accelerating pace of developments. Such developments include increased storage capacity, increased processing power and increased bandwidth. Since it is best to leave these developments to expert analysis, they will not be addressed here. I would, however, like to point out that technological developments are also influenced by our requirements and expectations – in other words, our needs. The past has shown that some of the wildest predictions or dreams have come true – even if it took several decades. Vannevar Bush, for example, described his memex, which is regarded as the forerunner of hypertext and the World Wide Web, in 1945.

To influence technological development we therefore need clarity on our needs and how to satisfy these. This seems to be a crucial factor in the future of information. Numerous research projects have already been conducted on needs assessments – user needs, educational needs, consumer needs, financial needs – but somehow it is still difficult to strike a balance between what we think we need, how we express these needs and what will really help to solve our problems. Needs assessment has always been an important step in any systems planning; it will be even more so in the future.

This discussion takes as its point of departure how our needs and the aims we set to meet these can influence future developments. In other words, this discussion looks at how we use our needs and aims to influence the future of information.

2. How Can We Shape the Future of Information in the New Millennium?

On basis of the global society sketched in the introduction, we can set an aim for the new millennium. How about empowering the entire world population to gain access to information and to assimilate shared wisdom or knowledge such as answers to safety, health and housing problems as a first step and, at a later stage, information or raw data? This should be done on such a level that individuals can gradually add to knowledge generation to first benefit themselves in their immediate environments and gradually also their wider environments. To achieve such an aim, we will first have to consider the following:

- satisfaction of basic human needs
- provision of access to an information infrastructure
- acquirement of required literacy levels

2.1. Satisfaction of Basic Human Needs

In his well-known self-actualisation theory Maslow argued that certain basic human needs must first be met before people will attempt to satisfy their need for social acceptance, respect and appreciation and finally self-actualisation. Human needs will affect information needs.

In very poor and underdeveloped societies information requirements are very limited and have mainly to do with the satisfaction of basic needs for safety, protection, food, housing and health care. Shared wisdom instead of raw data or information is required. Apart from food, shelter and health care, people need answers to their problems and not just information. They need expert advice. People in this category probably lack techno-logical access and literacy skills. Only once basic needs have been taken care of can these be addressed. Taking care of these basic human needs is the task of governments and influential components of society. This is an urgent task, since a very large part of society falls into this category – but it is also a task requiring substantial economic input and human resources.

Taking care of needs in this category requires expert advisers who under-stand the needs and know how to communicate with the people. Oral cultures and illiteracy may also impact on how information is shared and disseminated in this category and on the type of information systems and methods of information dissemination that need to be developed. These in turn will impact on future training for information dissemina-tion and systems design. Once basic needs have been satisfied, people's needs may move up into the higher categories, where needs include gain-ing social acceptance and peer respect. Information needs in this category may concern jobs, studies, business, entertainment and leisure. The final category is that of self-actualisation, where information and knowledge are core components. It is only in the latter two categories that people can really start to contribute to public knowledge. Everything indicates that there will be a greater demand for knowledge generation and for captur-ing such knowledge. Although the needs of these categories will also be an important part of the future, they are not be discussed here in any further detail.

2.2. Provision of Access to an Information Infrastructure

Numerous information resources are already available. The growing trend is in electronic resources which require specialised technological and communication infrastructures. Without these, underdeveloped sectors of society are cut off from the benefits information offers. Computer networks and telecommunication infrastructures are currently being introduced in developing countries at an accelerating pace. But there are also many societies which lack even electricity and which would first require power supply infrastructures. They also lack personal computers, satellite communication et cetera.

Without appropriate infrastructures, there will be an ever-widening gap between the 'information rich' and the 'information poor'. Such infrastructures are already available, and are constantly improving, as explained in the introduction. A limiting factor, however, is funds to implement them. Access alone is not enough. The satisfaction of basic human needs and the literacy levels required to deal with information and to use these infrastructures also have to be addressed.

Information infrastructures will help overcome geographical barriers – an aspect which will become more important as people move around – even into space. Although access to information will be increased, it will also become necessary to train people how to cope in this global information space or cyberspace. They will need to know about the social procedures, the netiquette, how to deal with new ideologies, how to access quality, how to distinguish between the 'good' and the 'bad'. They will also need to know about cultural differences, colloquial use of terminology and so on. A new field of study might arise from this, as might a new field of study of individuals' awareness of themselves in cyberspace.

2.3. Acquirement of Required Literacy Levels

To deal with information in the new millennium, different kinds of literacy will be required, ranging from basic reading and writing skills to information, computer, media, network and visual literacy (not necessarily acquired in this order – it appears that children acquire computer and media literacy skills long before they master reading and writing skills.) A growing emphasis on training courses in these skills and effective training and assessment methods will follow. For information literacy skills the emphasis will fall on the critical selection of information resources, the filtering, interpretation and application of information and ultimately knowledge generation. It remains, however, to be seen how effective these will be. In affluent communities children are, for example, encouraged to read from an early age. However, many enter adulthood with doubtful reading skills and little appreciation for information sources. When addressing the different literacy skills, our efforts should therefore not focus only on the methods, but also on how these skills influence human behaviour and attitudes towards information. Given the large population who require literacy skills, independent training methods such as computer-assisted instruction and online teaching will feature very strongly. Much can also be gained from the fields of distance education and adult learning.

Because of the large numbers involved, it may also be necessary to share responsibility for literacy programmes. Another issue that arises is the role of the English language in the future of information. Some search engines make provision for translations and for selecting the language of the interface. However, English is most often the language of interaction in finding information and in which information is presented.

In addition to the needs arising from empowering the global population to become information literate, there are also needs arising from current trends which should be addressed simultaneously.

3. Needs Following from Some Future Trends

The future is normally characterised by aspects that can be predicted with relative certainty on the basis of current trends. Many such trends can be identified in the future of information. There is, for example, the ongoing theoretical debate about the concepts of information, knowledge and expertise, as well as the growing awareness amongst ordinary people of information and its value. People do not doubt its significance: information is something you need to make a decision, to take action, to react appropriately, to protect yourself, to reduce uncertainty, to place yourself in a better position et cetera. Access to information has actually been proclaimed a basic human right and has been accepted as such by some countries. This could lead to demands for free access and maybe even free advice on certain categories of information, for example basic health, legal and educational information. It will also raise ethical issues about who is to decide what information will be freely available and who is to provide for the information and infrastructures.

There is also a growing need to provide access to electronic information resources. Internet search engines will continue to improve, but there will also be a bigger demand for resources selected, categorised, and described by people. Subject directories, intranets and extranets will become more important. So will personal home pages, replacing the long called for personal databases. All this will require time, expertise and often financial funding – which again will impact on equal information access.

Uniform methods to describe information resources in terms of physical features and content will become more important. So will the issue of vocabulary. Although automatic methods will feature strongly, there will also be greater reliance on what we have learned from thesaurus construction, semantic networks and so forth.

There is no doubt that the future will bring more and more information – often of dubious quality. Better means of dealing with it will be required. These might range from intelligent agents, expert systems and critical human analysis. What are 'quality' and 'appropriateness'? Much has been done in this field, but our information systems for and even human ability in deciding on quality still leave much to be desired. If quality assessment is left to technology and expert systems, we will need a better understanding of how we determine quality (many criteria have already been proposed) and also an understanding of why human judgement continues to fail. We will also need (expert) resources to check authoritativeness and the credentials of the creators of information sources. The more information we collect on people, their credentials, preferences, interests, income et cetera, the greater the need for data security and data protection. Data mining and data warehousing are important methods in marketing, but how do they affect individual privacy? Given human nature, chances of information abuse and malpractices are significant, which brings us to the danger of turning information into a powerful and negative weapon (e.g. for spreading pornographic information, dangerous ideologies, etc).The future will see a growing concern about these aspects and how to deal with it. (On the other hand, the publication of the first bibles was also marked by anxiety.) It is, however, evident that the evaluation of information, distinguishing between good and bad, will be emphasised to a greater extent in future literacy courses.

Information overload and information anxiety have been spoken about for many decades. More recently there has been talk about infoglut and information stress-related diseases. The situation is deteriorating with the growing ease of access to information and methods for automatically receiving information. As our access to information increases, our need to be protected against unwanted and unnecessary information grows. Some sectors of society can afford to pay for human intervention such as analysts, information gatekeepers and information specialists. This seems to be a growing trend, leading to a greater demand for people with appropriate skills.

Current awareness services which keep people up-to-date on new developments are the order of the day. Because of the overload, many people are, however, rejecting these resources: they prefer well-known, reliable human contacts. It therefore seems that we may see a move back to invisible colleges. The fear has also been expressed that the information overload will cause people to reject published research and focus solely on their own efforts. Apart from research on information organisation, information management and knowledge management, the future will also see research on and training in how to deal with information anxiety. The medical and psychiatric professions may find themselves treating diseases related to information stress. Diseases caused by working with computers, visual display units et cetera have already been reported. Imagine receiving doctor's orders to stay away from all forms of information for at least a month!

With unemployment and redeployment, there will be a growing emphasis on job performance, quality assessment, the appropriate use of information and trend analysis. Analysed, value-added and repackaged information will become very important – but probably also expensive. Life-long learning and building up portfolios of skills will become essential in the job market – a trend which is already detectable. This will require more educational material aimed at the independent and adult learner. The only difference may be that people may have more time to work on these because of shorter working hours to give more people the opportunity to be employed.

The timely delivery of information will become essential. The technology is already in place for this (e.g. electronic document delivery systems). These services are, however, also very expensive – thus adding to the inequality of access to information.

Our means for finding, analysing and repackaging information will need further research and refinement, for example data mining and data warehousing. Currently our information systems are based on the principle that information is essential for appropriate decision making. To improve systems we will require more research on exactly how and to what extent information is used in decision making. Information is already an important component of our consumer, leisure and entertainment decisions. Apart from our academic and occupational lives, it is also starting to play a major role in our personal lives. A move towards information literate societies will only add to the importance of information. It will, however, also add to the need to deal with the negative issues and to prepare people to deal with these.

Conclusion

When considering the trends mentioned here, it is clear that the future of information will be characterised by growing urgencies: quicker access to information; faster information retrieval; better quality information; more relevant information only; cheaper or free information; better competitive edges; better job performance; improved knowledge and expertise; and better capturing of knowledge and expertise. Research in Library and Information Science on subject analysis, subject description, user and search behaviour, user needs and the use of expert systems and automatic methods has already addressed some of these issues. We will, however, require improved capability to apply our knowledge and skills, and to refine our research methods. In other words, we will have to be better at what we are already doing.

Our anxieties about coping with information, assimilating it and adding to knowledge generation and gaining a better understanding of these will also grow. So will feelings of inadequacy. Depending on how we deal with these, and also with other needs that have been identified, the future of information may see a growing dependency on information and information technology, or a return to reliable human networks of information. The future of information will rely on our ability to identify our needs and to express these in terms of the technological developments required. At the same time it will depend on whether we accept responsibility for empowering the global society to become information literate and to add to knowledge generation. Our success and the ultimate direction of information will, however, also depend on how we address the urgencies and anxieties associated with information. The ability to address these, and to adapt to changing circumstances, should form core elements of the training of information specialists and of training in information literacy.

33

The Tail Wags the Dog: The Future of Information is Now

Dave Nicholas and Tom Dobrowolski

Introduction

There is greater justification for thinking about the future of information as thinking about the future of anything else (the environment, for instance). In the fast-changing and hothouse world that information inhabits, information forecasts, visions, musings etc. are soon proved right or wrong – certainly within the memory span of the audience or readership. The soothsayers amongst us – and there are many, as this book testifies – are in the unique, if not unenviable, position of prophets who live to see the truths and falsehoods of their own prophecies unfold. This of course means that we can – and should – really learn from our mistakes.

By making forecasts of the future we are in fact making statements about the present, because the whole process starts with an appraisal of the present, a look at the fundamentals and the lie of the ground. There are many examples of bright, right visions of the future, like those of Vannevar Bush, J. C. R. Licklider, Ted Nelson, Marshall McLuhan, Arthur C. Clarke, and Tim Berners-Lee. Their visions have had a powerful influence on the development of information systems and structures. However, they did not just predict the future (correctly) but they also helped create the future they envisaged. Nobody knows in what direction developments in information services would have gone if their forecasts had not been made. By making forecasts now the future is already being mapped out in a certain kind of way. Prediction and creation are intertwined. Thus it was McLuhan, who by observing the present, made one of the most amazing visions of the information future. While observing the television of his day he went on to describe the web. The right vision of the future shakes reality and starts to shape it in new ways. The tail wags the dog.

But what do we really mean when we use the term 'future'? What time period should be considered. A decade? Twenty five years? A century? Well, it would probably be impossible to see beyond a decade when it comes to information's future. Thus the British Library's attempt at forecasting the future, entitled *Information UK 2000,* and published in 1990, spectacularly failed when it failed to mention the Internet. It can be done, though, as Arthur C. Clarke showed only too well in *2001: A Space Odyssey.* Arthur C Clarke went to the same school – Huish's Grammar School, Taunton – as the lead author, so we are feeling brave. Four concepts/ issues/phenomena will dominate our thinking over the next decade – 'dumbing down', 'the tail wags the dog' (i.e. information systems lead rather than follow), individualisation/customisation and biotechnology.

The Forecasts

First lets consider what it means when we use the phrase 'the future of information'? Is it a synonym or near synonym of the 'future of information systems'? Where and what is the difference? In fact, it is much easier to consider the future of information systems than the future of information. In its old medieval meaning information was a synonym for knowledge. In its modern usage one can witness a reduction in the meaning of this term – a 'dumbing down' if you like, because now information is a synonym or near synonym for data. However, the old meaning still exists. Information science in its theoretical deliberations does still talk about information as a synonym for knowledge representation or knowledge presentation, but in practice, particularly in the field of information systems, information means data. There is a tension between the two basic meanings of information: information as knowledge, and information as data. Data is the commodity and concern of information systems; knowledge is much more personal, the domain of the individual. The definitional problem is strictly connected with the 'dumbing down' process that is the real and inevitable consequence of the social flow of information. The mass consumption of information requires that the message is simplified. The distribution of information via a mass media system inevitably results in the loss of meaning. Knowledge becomes data. Why? Because information systems are only loosely connected with personal knowledge. Because it is a part of an information system, unconnected with personal values. Compare that to the way our memory works – it reduces information, or strictly speaking, selects and filters information. Only information connected with personal values, individual points of view, philosophy of life, is processed. All the rest is abandoned to the round filing cabinet. Information systems process data; people process knowledge.

It is much easier to say something about the future of information systems than about the future of knowledge systems. The really fundamental question is how closely is the future of science connected with documents. Is it possible, for instance, to create theories and laws without explanation, without descriptions of the matter? What does it mean when we describe something as an online document? Essentially, we have two types of documents: genuine documents and quasi-documents. Quasi-documents are usually created as a result of a personal online search. Who is the author of that sort of document? The online system? The database editor? The user? We believe that the semantic net of documents and quasi-documents will be much more extensive in the future and searching for inspiration will be easier than now. We believe also that we can say something about the future of knowledge only by speculation about the future of information systems.

The real problem is how to facilitate the serendipitous location of information. The semantic net of documents facilitates the association of facts, stories, ideas, data. If you want to find data, search by data. If you want to find text, search by text. But if you want to find documents, search similar documents as your preferred ones. The document is much more than data, text, images and another objects connected all together. Documents describe a way of thinking, a way of our understanding a problem. Looking for similar documents is the best way for finding patterns for our thoughts.

Predicting the future by extrapolating from existing statistics leads to lots of mistakes. And that was the problem with *Information UK 2000*. The better approach is to look for quality of life markers or milestones that need to be passed. If you want to say something about the future, say something about quality of life and the possibilities for its improvement. The web is changing almost daily – could you find a better symbol of the future, a better auger of things to come? The future of information is with us now. One can expect faster transmission and free access to the Net for all, 3D interface of Windows and WWW, and much else, but the key components of the future are here.

Presentation and packaging has become as an important part of a message as the text. The interface is a part of that presentation. The interface is information because it is tied to a spectrum of different styles of presentation and can be substituted for other styles of presentation. The key word is autonomisation: presentation obtains autonomy (independence) from the text. Texts are connected not only by their meaning but also by the way its presented. The chosen style of presentation becomes information itself. The medium is the message in the world of WWW. The process of autonomisation of presentation (i.e. independence from text meaning) is another manifestation of the dumbing down of information. The text ends up in the background, and presentation in the foreground. Presentation in no longer tied to the meaning of the text; the difference between high culture and low culture does not exist in cybermedia. The same interface and presentation tools might be used to publish Shakespeare's sonnets and the weather forecast.

One can observe the same process happening to searching facilities. The search itself becomes the payback, not, as once was, the resulting information. Searching becomes a recreational and leisure activity and it has been conveniently rebadged as 'surfing'. We are looking for entertainment, a show; we want to sail in the ocean of information for an intellectual adventure. We are also looking for motivation, money and pleasure. We are also looking for a creative atmosphere. Full three-dimensional interfaces can only strengthen this behaviour. Navigators of the future will be much more emotionally and intellectually dependent on the Net than we are. But the dumbing down process also has a creative face to it. Information remakes and sequels are much easier to make if the process of dumbing down has been completed. Dumbing down of the text means it has been simplified and standardised. Standardisation is a core of any technology. You can create new shapes from standard bricks. There is connection between standardisation and the dumbing down of media and the standardisation (and packaging) of our needs and imagination. This is the influence of technology on our way of thinking and feeling.

The shrinkage of space and time, which seems to be the hallmark of the most successful of all information and communication systems can also lead to the dumbing down of documents. Hypertext documents are shorter than traditional articles and remarkably shorter than books. There is no space for detailed description in hypertext documents. These 'fragments' of text are bolstered by links to genuine documents. Writing HTML documents becomes all about knitting links to other documents and multimedia objects. If this observation is right, future intellectual life will be much more fragmented than at present. Those who need a big canvas to express their ideas, opinions and philosophies need to write a book. Printed books are not only media of information but also a conventional means for the expression of ideas, opinions, and theses.

It is impossible to say something about the future without mentioning the web lots and lots of times. The World Wide Web is a unique example of right vision (hypertext), right technology (the Internet), and huge publicity used together to create a powerful system of information. The WWW is based on open systems philosophy. This idea might be understood on many different levels but whatever the level it will undoubtedly be a base for any future system. The open systems philosophy is very much connected with the intuitive interface and a vision of online systems as public media systems.

Plainly the world of work is full of 'Mac-jobs' and it is possible that future systems could go down this route too. But what the information past has told us is that things never turn out to be quite so straightforward, quite so black and white as first envisaged. The reverse of dumbing down could in fact happen. And if it does it will all be connected to personal knowledge-based information systems. A dumbed-down version of information will be available for the general public; complex information will be for personal use. Creative activities require complex information. Try to be a gardener or fisherman without extensive information about your garden or fish. For this to happen we will first require user-oriented services based on personal knowledge.

The information society will never truly take hold until we can genuinely meet people's information needs. We talk glibly about the commodification of information but essentially we still very much crude batch processors. Information products are incredibly raw, people get by with imprecise or no information at all – and at a cost of great inefficiency to their lives and work. There is a mistaken belief that the future is all about sharing information – knowledge management style – or storing and distributing information – digital library style – but it is in fact about getting closer to what people need in the way of information and producing it in a processed and packaged manner so that the individual can consume it at any particular point in time, any location. If we want to do for information what fast food has done for the catering industry then we will have to do this. But customisation, individualisation, segmentation – whatever you call it – can only come on the back of personal detail and knowledge. The logs, subscriber databases, the cookies and the rest of the cyberspace police provide us with the means by which this data can be collected. It would be incedibly foolhardy to deny the dash for personalisation in all things information, but while the industry generally acknowledges the inevitability of the move to personalised information they seem to be doing precious little about it. But when they inevitably do, the tail will wag the dog, and how.

It is difficult to predict whether the huge publicity that attends every move of the web and proclaims it to be the most remarkable information system of our time will pass on to any of its successors. It must happen that the Net and connected technologies will be taken for granted and will end up forming a basic – and untrumpeted – part of infrastructure, like the electric supply. As the most commented technology its place will surely be taken by biotechnology because of the dramatic consequences of applying it. The biotechnology revolution will have even more profound consequences than the information revolution although its power will have been enhanced and amplified by it. A synergy between both technologies is possible (biocomputers). We shall then, undoubtedly, be at the crossroads of Western civilisation. What repercussions will flow from the genetic modification of mankind? How far can we modify our natural biological environment? These questions are so profound, so serious that they will come to the fore of any public considerations. Revolution is very difficult to control, and ironically even more so now that we have the Net. Development of biotechnology will be very difficult to control and so powerful that the twenty first century could be more volatile and violent than the past one. Public opinion will surely be the main stabilising force and control of biotechnology development. Because of the Net and information technology, public opinion will be more powerful than ever. A non-global solution to acceptable development of biotechnology might lead to the foundation of very different civilisations. The differences (moral, legal, social) among them might be more profound than those that existed between Athens and Sparta The biggest challenge in the future will be acceptable boundaries of biotechnology development similar to those created to control nuclear and chemical weapons. And the web will have a role to play here.

Conclusions

We can confidently make three major assertions about the future, that:

1. any new and revolutionary information systems will be global Net based
2. information systems will be much, much more user oriented
3. the information economy is unlikely ever to overheat.

We can add that:

- creating new information needs will be more important than meeting existing ones
- many more services will be created to satisfy personal knowledge needs – we have just started along this road
- looking for personal inspiration will become more important than looking for what is now called information
- information professionals will become a part of the user's team, and no longer part of the provider's or system's team
- it will not be data mining but personal knowledge mining that will be the key to the future. Personal hypertext documents will include our intellectual preferences, and comprehensive description of information need. The future searching of the Net will be based on looking for similar documents as our personal hypertext documents
- the study of information needs will then inevitably become the core of information science, and should, as a result, send the pendulum the way of the information professional

- the production of information and information systems will not reach saturation point, because new services will replace the old ones before saturation takes hold. And the flow of information really means the consumption of information. An overflow of information is needed because information is constantly being consumed. It is very similar to what already pertains in the media world – and we are ever drawing closer to that world. The information economy does not operate like the classic economy that is constantly trying to balance needs and supply. The balance between needs and supply is much harder to reach in the new digital information and communication economy because new needs are created faster than information is supplied. The best known example of creating new needs is mobile telephony. Most users of mobile telephones have conventional telephones already, but that hasn't stopped them fuelling a massive expansion of the mobile telephone market.

New communication and information services create new patterns of modern culture. The next culture to dawn is surely that of the individual – and it is only the information developments in the pipeline that will make this possible. The tail wags the dog so hard and painfully that the tail dictates the route the dog will take – and the signs of this happening are already with us now. The future is already with us.

34

Marchers in Time

Barbara Quint

Toward the end of the first century of the Second Millennium, a group of English scholars and bureaucrats worked on a memento for the time called the Domesday Book. Here we are on the edge of the Third Millennium (give or take a week or two), a group of English-speaking men and women (see globalisation, see emancipation) joined in another 'doomsday' venture: predicting the future, not just for a mere century but for a millennium. Suicide pills in our pockets, our happy band approaches the task clinging to the knowledge that, unless medical researchers have grossly underestimated their success rate, at least we'll be long gone before anyone starts sneering at our ridiculous misperceptions.

How can we predict the course of the march of time? Well, first, we should probably watch the marchers. They say that no man is a hero to his valet. Well, no researcher is a hero to their research librarian. In the decades I worked as head of reference at a major policy research organisation, I saw research techniques and methodologies from a broad array of disciplines applied to detecting outcomes of alternative policy options. Some worked rather well, others didn't. The only predictor that always worked, one way or another, was demographics. People are born, grow, work, mate, breed, sicken, and die, not necessarily in that order.

Some would say that all major trends and developments in human history can be traced to underlying demographic realities. Were there really two World Wars? Or did the horrific casualty rate of World War I simply require 20 years to breed more warriors for the second round? Belief in demographic forces can become quite mystic. In *War and Peace*, Tolstoy posited that the French invasion of Russia was an inevitable, lemming-like movement of populations, completely independent of the will of Napoleon. Regardless of the intensity of one's belief in demographic forces, one should never ignore them. They may not determine success, but defying them will certainly insure failure.

In the first quarter century of the new millennium, the vast swell of post-World War II population that has shaped the history of the last half-century, the Baby Boom generation, will grow old and die. Traumatic and stressful as this inevitable reality will be to individuals and societies, it could have a positive effect on information technology and services.

(Every cloud has a silver-haired lining?) As they age and weaken, the Boomers along with their money and resources, will look for products and services that make their lives easier. With their declining mobility, they will want services delivered wherever they are. Telecommuting is the beginning of the process. Distance learning could play a role. E-commerce is an inevitable winner. (If you play the stock market, bet the pension money on UPS and Federal Express.)

Though the Boomers may drive such developments, they will not contain them. Their children and their children's children and their children's grandchildren will have been raised in a world where 'fetching' may be an option, but never a necessity. Think how that can change things.

Think of the changes it will impose on institutions that have defined themselves by place features. One group comes to mind: librarians. We will have to burst out of the buildings that bind us if we plan to prevail in the new information world order. A good first step might be to rename our professional organisations to accurately reflect that the members wear clothes, not landscaping, that they eat food, not interior decorations. After all, a library is just what's left over when the librarian goes home at night like a coral reef just represents the activity of living coral.

Of course, demographics are not the only forces in play. Technological developments clearly affect outcome. The juggernaut of computerisation will continue to roll across the globe.

Anything that increases ease of use will prevail. Already we see the movement toward the flexibility of flat panels and liberation from wiring. As prices come down (another inevitability), we will soon reach the breakthrough point for book readers of 'belly-compatibility,' i.e. a cheap, easily viewable reading device that one can use while lying prone, elbows propped around a supporting pillow, head cradled in hands.

The next major break point for user-friendliness will undoubtedly be fully effective voice input. At that point the computer will have achieved the ultimate data input format – human speech. Combine this development with full multi-lingual translation and improved world communications and we will have empowered globalisation at a whole new personal level.

If you step back and look at the course of computer development, you see two constant trends. Behind the scenes, engineers and super-techies work continuously to increase the technology's efficiency, reliability, and power. Up front, however, we see the computer become more and more human. No, we do not refer to the all too human capacity for error as expressed in the Y2K controversy. Although one does wonder on what grounds the U.S. Congress could justify an anticipatory bail-out bill for litigation stemming from Y2K problems ('Honestly, judge, it isn't fair. How could anyone expect engineers with Ph.D.s to know some other number came after one thousand nine hundred and ninety-nine?!'). We refer to the computer as a humanised and humanising force. Computerphobes and writers running dry often ridicule or castigate the nerd herd whose only human contacts take place online. But 'let us not to the marriage of true minds admit impediments' (how's that for a literary bundle for Britain?). Anyone who has used email, listservs, newsgroups, chats, or other virtual community tools knows that they represent a wonderful expansion of human outreach. As a byproduct, they have also revived the art of writing as a popular skill. Project the humanising force of automation onto a global scale after voice input and machine translation have triumphed and you have a person-to-person force that the world will have to reckon with.

All cutting-edge technologies can cut two ways. One reality that will distinguish the next millennium is the loss of forgetting. Imagine living in a world that cannot forget, that remembers and stores everything – every image, every word, every document, every joke, every insult, every message, every call. Are you cringing? Nor can anyone stop the development with policy rulings. That Boomer generation of which we spoke will need universal, intrusive, automated memory to get it through the terrors of Short Term Memory Loss. Once the data, no matter how personal, is accrued, expect it to be shared; in fact, the Boomers will need the data shared. They will have to rely on vendors to re-supply their kitchens and bathroom cabinets as their memories fade and harried relatives look for shortcuts in elder care. Once people have developed the habit of relying on computers as memory substitutes, they will never turn back.

Content once accrued will circulate. Look at the cable television if you don't believe me – all those channels channeling any content they can acquire over and over again. With PC-TV convergence on the way, one can look forward to services that let people pick their decade. A simple scheduling choice and you can spend the next month in the 1950s or the next year in the 1930s.

In fact, old content could create new content. Multimedia developments and advanced computer animation will produce new material from old. Ever wondered whether Vivian Leigh's Scarlett O'Hara got back together with Clark Gable's Rhett Butler? Someone could write a sequel and use computer-generated images of the two stars, as they would have appeared in their own time, to generate new performances. As salaries rise for non-virtual leading actors, the price of virtual established stars may become very reasonable. Wonder who would own the copyright? The studios that supplied the images or the estates of the actors?

With full Internet and web connections, people could pick their virtual realities. Vendors won't mind as long as they can make a sale. In late 1999, the *Wall Street Journal* (http://www.wsj.com) ran an article on

Microsoft's Encarta encyclopaedia which revealed that Microsoft offered different facts for different cultures. For example, an underappreciated Italian-American invented the telephone in the Italian language edition, while Alexander Graham Bell continued to prevail in English-language and German editions. Will non-virtual societies begin to disintegrate as individuals select their realities, define their peers, and commit to new loyalties? To quote Bette Davis in *All About Eve* (or Joseph L. Mankiewicz, if you believe the Screen Writers Guild version), 'Fasten your seat belts. It's going to be a bumpy night.'

In all this confusion, in all this unrest, in all this chaos coming upon us, who will help people everywhere protect themselves from the dangers of inaccurate, overpriced, dangerous data, from the perils of relying on the unreliable and sinking into virtually baseless virtual realities?

Librarians, that's who! Or information professionals, as we may call ourselves. As the old, but still painful, joke goes, 'That's why they pay us the big bucks.' But really, who else? Of all the information professions, librarians are the only ones that define and measure performance solely in terms of clients' needs and interests. All the other professions define their members in terms of technical platforms or vendor interests first and clients second, if that. Programmers only deal with problems that programs can solve. Information industry professionals only develop products that can produce a profit. Only librarians use any medium that works to gather any data that is true to serve the client. For librarians, it ain't over till the client is happy. We work for smiles.

By the way, as many Net newbie firms have discovered, such client-focused professionals make marvellous managers and employees for firms truly committed to the interests of users. However, librarians will not be in a position to do for their global clientele what their global clientele needs done until they themselves design the products and services, until they become the Net newbies supplying virtuous, virtual library service. Librarians have to build services that serve and create global constituencies. They cannot confine themselves or cling to the geographical limits of politically or financially defined constituencies, such as those of a government or corporation. Information technology has made and will make such limits seem antediluvian in the Third Millennium. In any case, the clients that need us most are, more and more, out there on the web. We need to take the service to where the clients are.

So in a year or two, the web and its denizens will quiver as word of keyboard sweeps across the earth bearing the glad tidings, 'Help has come at last! The cavalry has arrived! Robin Hood and his Merry Persons have saved the land! Quick!! Change your start-up pages to http://www.thetruth.com (or www.thetruth.org for you purists).' As the decades pass, this mighty web site will stand like a beacon, guiding weary and frightened travellers to safe harbour. It will carry only audited, evaluated, confirmed sightings of accurate, unbiased (or bias-identified and cross-checked) data on approved sites. Its lucid interfaces will tie that data to real-world questions in customer terms. Its steadfast links will provide full documentation for the sites and register any tips or techniques needed to extract sound information, as well as any reservations or limitations of that data. It will incorporate the educational elements needed to enable a customer to understand and effectively use the data they extract. Most important, it will confirm when a source for data of the quality required does NOT exist on the web, but in another medium. It will then identify that source and that other medium and provide access routes.

In time, webmasters across the length and breadth of the web will beg for TheTruth.com's stamp of approval. Tamely, they will submit sites for TheTruth.com's eagle-eyed and noble-browed consideration, trembling in panic lest TheTruth.com condemn their offerings with its 'Liar Liar' anathema. With its massive user base, TheTruth.com will inventory all the data for which customers have expressed a serious interest, but for which the web lacked adequate sources. This inventory will serve two purposes. It will challenge other web citizens to prove TheTruth.com wrong and find a qualified source to their everlasting glory. Through the awards given to customers who have proved it wrong, TheTruth.com will confirm the purity of its commitment to service, that it never lets ego interfere with its professional service ethic. The inventory will also comprise an invaluable source of revenue which TheTruth.com can sell to the information industry to inspire them to fill the gaps in the web's information base. 'Do good and do well' is the motto of TheTruth.com

Who will create TheTruth.com? Who will craft this mighty force for good? Librarians of the World united to serve. They will join together in their professional organisations and hammer out rigorous standards. They will establish virtual consortia that link committed web librarians together in a global network of quality-minded seekers for TheTruth.com. They will bring the profession undying honour and protect a world of clients from a world of harm. And in case you worry that TheTruth.com has grown too rigid and establishment-oriented in its definition of truth, the New Age, X-File generation among you can work on their alternative service YourTruth.org (or YourTruth.com, for you realists).

Am I dreaming? Yes, but it's the right kind of dream. Above all, information service in the new millennium will involve radical changes and radical new organisations to create and deal with the shifts. No more incremental advances, only leaps of faith.

Let me offer one word of advice and one word of comfort for the scary times that have come upon us. My advice? As you face the challenge of this new time and scan the horizon for signs as to the new directions it will take, focus your attention on underlying structural requirements. Listen for those clicks as key elements lock into place. For example, take distance learning. In the spring of 1999, one university, the International Jones University, achieved full accreditation from one of the regional agencies that accredits all the colleges and universities in the United States. Click. Within weeks, the Open University in the U.K. began global operations. Click. Columbia University announced that they would open up an office to partner with vendors seeking to establish distance learning operations. Click. A leading academic publisher announced that it would contract with professors to establish courses and curricula built around their textbooks. Click. The U.S. Department of Education announced that it had begun establishing student loan criteria for students taking classes from 117 distance learning educational establishments. Click. Click. Click.

Apply the same research technique to other areas moving forward rapidly: e-commerce, health information, etc. Analyse the underlying structural elements that make an existing operation work. Estimate what advantages digitalisation of the process could offer, what hurdles such digitalisation would have to leap. Figure what elements and components would have to be in place for the switch to occur. Sometimes the sound of the clicks can be very quiet. Sometimes they can be muffled, but they're there. And who better to ferret them out than information professionals?

And as for that word of comfort, remember that the world we have spent our blood, sweat, toil, and tears upon has finally arrived. We were online before online was online. We cared and nurtured information services before they turned a profit. As more and more everyday people become information handlers using the technologies we supported in their infancies, they will need professional guidance and support. They will need protection from people who only view information as a way to make a profit or as a tool to control the suckers. We librarians, we information professionals, belong in the Third Millennium because the people of this new time need us now more than ever.

The Third Millennium has arrived. We're home at last.

35

Chinese Web

Kevin McQueen

I have seen the Future, and it is Murder (Leonard Cohen)

Deep in the heart of the 2nd Tantou Industrial Zone in the Songgang district of Shenzhen, the Industrial Revolution has hit China on the cusp of the 21st century. Just across the border from Hong Kong, the zone looks more like a scene from an apocalyptic road movie – dusty roads, nary a hint of greenery, wild-eyed inhabitants, buses that look like they've been in use since 1949, and unique home-made vehicles that seem to be part Harley-Davidson, part Panzer tank and part golf cart. Mad Max would be at home in this place! But unlike Mad Max, the people of southern China are not living in isolation. There are swarms of people everywhere you look, and on the congested highways you can see them darting across the road between the dilapidated vehicles and through a haze of exhaust and smoke.

The sheer scale of development in China is so overwhelming, and the attendant greed and corruption so mind-boggling, that it is hard to resist comparisons to Hobbes' State of Nature – an anarchic netherworld of all against all. But, of course, the reality is much more complex. China is trying to play catch up with the developed Western world, and damn the consequences of 'progress'.

'But what does all this have to do with information?', you might be asking yourself right about now. Well, I can't speak for information professionals, but as a journalist I am an information gatherer and there is no doubt that information is power. It plays a large part in politically managing the people of China as they are dragged kicking and screaming into the 21st century. The fact is that while the Chinese recognise the need to modernise economically, they fear the consequences of full-scale Western liberalism. They fear the consequences of the masses gaining access to information and losing their power, thus they seek to control that information. While you and I may take a relatively free society for granted, despite the increasing reach of multinational corporations, the exact opposite is true in China. There, information – and the whole gamut of technological and economic advances that make up progress in the West

– are used to control the population as much as to educate and enlighten. Nothing is more potent a symbol of this than the Internet – the Chinese need it to help disseminate and process information, but they fear the impact of too many outside influences. I had first hand experience of this in early 1997.

I was a reporter for a local Hong Kong business magazine when I first stumbled across a press release from public relations firm Euan Barty and Associates on behalf of California-based computing network firm Bay Networks informing me of a soon-to-be launched web site, the China Wide Web (CWW). January 1997 was an auspicious date, given that Hong Kong was going to be handed back to China from Britain in a little over six months time. The faxed press release read as follows: 'Bay Networks Inc. has been chosen to provide the infrastructure for one of the most ambitious and far reaching information technology projects in the history of the People's Republic of China – the China Wide Web.'

Bay Networks had been chosen to design and implement the China Wide Web in a multimillion US dollar 'technological partnership' with the China Internet Corporation (CIC). Owned by Xinhua (the New China News Agency) and Hong Kong investors, the Hong Kong Internet service and content server planned to expand its servers through China. Based on a network of powerful servers located in over 50 cities across China, the first five key sites targeted were based in Hong Kong, Beijing, Shanghai, Shenyang, and Guangzhou. It was billed as to be primarily used by mainland companies, joint ventures, and government agencies. At the time, private citizens were still not allowed to register for Internet services, which went hand in hand with China's tight policing of the Internet. Only the sanctioned elites of the business, government and academia were allowed access.

The set-up of a national Internet and computer infrastructure was very much in keeping with China's continuing and insatiable drive for economic modernisation. It has all come with the blessing of the Communist authorities, and the decision to put Xinhua at the fore was no accident. Xinhua is the official information arm of the Chinese state, and in the period leading up to the handover of Hong Kong it was seen as the de facto Chinese embassy. There was no way that CIC could just merely be another competing Internet Service Provider, nor even just another mainland firm with *guanxi* (which loosely translates as 'connections'). I understood very well that the site was largely benign in its content; it was to carry a lot of business news useful to a lot of investors looking to pump money into China in the hopes of catching a slice of its burgeoning market. It was between the lines of the press release that the other more ominous intentions lay.

'CIC is utilizing the latest Internet technologies to build a highly reliable and secure national intranet.... The CWW's unique intranet model provides business information while addressing the needs and situation in China....The CWW network will be based on Xinhua's existing satellite network backbone linking the major cities and provincial capitals of China.... Established and operated by CIC, CWW is China's first nation-wide intranet (its name in Chinese means the 'web inside the country').... While the World Wide Web (WWW) has exploded into prominence in the Western world as the new medium for electronic communication, commerce, and information retrieval, language disparity and legal sensitivities hamper the WWW's widespread growth in China.'

It was obvious to me that CIC would control and censor content through their network, but equally obviously they didn't want to publicise this too loudly. Backed by Hong Kong and Western investors, CIC knew that too much bad press in a Hong Kong that at that time was frankly jittery about the handover, could be harmful The previous month had seen the spectacle of a Beijing-backed 'shadow' legislature being sworn in across the border in Shenzhen as China's relations with Governor Chris Patten had gone from bad to worse. What was so striking to me about Bay Network's involvement at the time was the willingness of an US company to put up with this sort of deal with a Chinese government agency. But I guess money always talks, and in hindsight I was naive to be even mildly surprised. Many Western and Japanese companies are looking for riches in the Chinese market, and fail to see anything but the promise of filthy lucre: 'technological partnership' being their ethical loophole. Rupert Murdoch once boasted that his satellites would be the bane of totalitarian regimes, but when the Chinese government warned him that they would pull the plug if his Hong Kong-based Star TV network did not refrain from beaming the BBC World Service into the mainland, he complied.

At the CIC/Bay Networks Press Conference held at Pacific Place in central Hong Kong on the14 January 1997, CIC and Bay Network officials skirted the issue. At the press conference were Michael McLeod, CIC (HK) director of sales and marketing; Bill Ting, Bay Networks' systems and engineering manager, Asia Pacific; Jeber Chu, Bay Networks' Greater China area manager; and Aaron Cheung, CIC (HK) vice president of network operations.

'We do not plan to compete with Internet service providers in China,' said McLeod. 'We see them more as partners.' He said that the aim of the China Wide Web was to help facilitate business communication links in China, with international gateway links to the outside world. Questions of content were dismissed as 'irrelevant' outside of the Chinese context.

'We have to make sure our customers get the right kind of information,' added Chu. This led to a number of the assembled press corps to ask questions about content and content control, but unsurprisingly it only produced a lot of huddling and whispering behind the microphones. No one was prepared to answer those questions, and even when I persisted afterwards and phoned Xinhua on the off chance that I might get a straight answer, a woman from Euan Barty soon rang back telling me that asking difficult questions wasn't such a good idea.

There is nothing wrong with setting up a network allowing businesses to communicate with one another in China, or giving them access to financial and business news from Xinhua, Bloomberg, Reuters, or from anywhere else. Ultimately, Xinhua can put whatever they like on their own site; the problem arose from the fact that my belief was that CIC would use the technology they gained to block what they deem to be objectionable sites. I was given encouragement by a column in the *Los Angeles Times* written by Gary Chapman that addressed some of those issues and confirmed my initial suspicions.

Chapman wrote that Bay Networks had provided the CIC with security and filtering technology managed by network firewalls to manage their content. The excuse usually given is a crackdown on pornography, but what really frightened the Chinese authorities were web sites on human rights, democracy activists, protests about the Chinese occupation of Tibet, and advocates of Taiwanese independence. All Internet users in China are required by law to register with the government, so even if you do log onto a banned site it isn't too hard for the authorities to check where you have been. If you get worried about having employers checking up on which sites you've been checking out, imagine how it must feel for the poor souls that find themselves under interrogation by the Chinese police.

I had eventually been able to get in touch with Bay Networks, and their marketing director, Annette MacDougall, made it clear that what Bay's relationship with CIC was a 'technological partnership' and not a 'business partnership'. I found that somewhat disingenuous as I was quite sure that Bay Networks had not exactly provided the backbone for the China Wide Web free of charge, but they did reiterate that questions of content should be directed to CIC. Well, I did that many times, and eventually an irate Michael McLeod called me up one afternoon saying that I had a serious misconception about what CIC was all about. He invited me around to 're-educate' me on the CIC. I thought that a spectacularly poor choice of words. On the day I was to meet him, he was taken ill quite suddenly.

At the time I got a bit excited by this story, and I thought that maybe I was being a bit sensationalistic. But my experiences in Hong Kong, China, Malaysia and Indonesia have taught me that information will be controlled in the interests of elites. By the time the handover came and went, I forgot about the story. Much of the concern died after it became clear that

China was going to let the local press carry on as before. Besides, China is well aware that Hong Kong is their window to the outside world, it is in effect the goose that has laid the golden egg.

Hong Kong – despite all the rhetoric about the imperialist British – is a godsend to modern Chinese aspirations. Modern, sophisticated, brimming with energy and entrepreneurial zeal, open to the outside world, there was simply no way that China was going to kill their golden goose by rattling too many sabres and frightening off investors. However, that does not mean they were any less successful in seeking control of the situation. It seems perhaps a simple analysis, but is nevertheless fundamentally true. Beijing has a handpicked Hong Kong Chief Executive in Tung Chee-hwa, a rigged electoral system, and a suitably compliant business sector. And there is subtle pressure on advertisers not to upset the apple cart too much. Publishing proprietors are chiefly interested in profit – à la Rupert Murdoch – and that ethos is passed onto editors and journalists. The problem of self-censorship starts in that journalists are kept away from certain stories before they are even out in the public domain. It is a real problem in Hong Kong for local publications – I know for a fact that it has happened in both the venerated English-language *South China Morning Post* and in the *Hong Kong Standard*; it is even more of a factor for the much larger Chinese press. It is less so for international magazines like *Asiaweek* or the *Far Eastern Economic Review*, but they have American backers and the option of running off to Singapore or even Sydney. But again, Rupert Murdoch is an instructive case. His publishing company, HarperCollins, refused to publish ex-Hong Kong Governor Chris Patten's book *East and West* for fear of offending Beijing. It isn't that old Rupert fears offending the Chinese, but that he doesn't like the idea of them cutting into his more-than-bulging wallet. For the most part, though, while the newspapers aggressively cover local politics, they are much more cautious about reporting the mainland. Hong Kong has a lot of very good journalists in both the Chinese and English-language press, but more than enough sycophantic publishers and writers to keep the propaganda flowing for some time to come. It seems the top-down control of information continues unabated.

But were my fears about the Internet in China justified? I believe they were, especially when you look at some of the developments since the handover of Hong Kong on July 1, 1997. Basically the same patterns have occurred despite the occasional tough talk emanating from the United States.

Western countries generally ignore the fact that China has a terrible human rights record, is likely engaged in genocidal policies in both Tibet and Xinjiang in the northwest, and is increasingly acting like a bully on the world stage vis-à-vis Taiwan. In terms of current international relations, Western countries – and by implication Western journalists – are told to stop interfering in internal Chinese affairs, and apparently that includes bringing in outside opinions and different ideas about the nature of life, death and politics. They then drag out the tired old cliché of Asian Values to beat you senseless with.

As Asia, and China in particular, modernises itself, they recognise that it will be increasingly difficult to control the amount of information flowing in, not only from the Internet, but also through television, radio, pop culture and foreign publications. But they do have ways of controlling it, and not merely by blocking web sites. In China you still have to register with the Government to get onto the web, and there have been stories in the press over the last few months of a crackdown on unlicensed Internet cafes in affluent Shanghai, as well as arrests of web dissidents. When you look at all the talk in the West of encryption codes and of companies starting to monitor their employees emails more and more carefully, again you can imagine what China can do with their state powers. There is a distinct danger that modern technologies in the West – the Internet, CCTV cameras, smart cards – will infringe on civil liberties. I will be bold enough to venture that in China these new technologies will be used to help them manage their society – the effect of the Internet may be to do exactly the opposite of what many Western pundits, gurus and charlatans expect, much less hope.

When I wrote my article in March 1997 there were only some 40,000 estimated Internet users in China; just two years later this figure is now estimated at some 4 million. It is burgeoning, but to expect it to be some sort of 'weapon of freedom' is naive to say the least. There is already a climate of fear built in with the Internet, even the idea that the authorities can electronically track you is daunting in itself. It does create a bit of a chilling effect, and the middle class and elite Chinese on the Net are not for the most part going to have too much interest in challenging the power of a state that has allowed them to achieve their wealth. As for the masses in China, they are a long way from being wired up, so the government will rely on more 'traditional' methods of keeping them in line.

As for CIC, it is interesting to see how they have grown. Besides the portal site of CWW – which does largely as advertised and only provides financial news through agencies like Reuters, Dun & Bradstreet, as well as Xinhua – three other more expansive portals have been set up for Greater China: China.com, Hong Kong.com and Taiwan.com. Despite huge losses since being set up in 1995, China.com (a sister company of CIC, which means they have essentially the same shareholders), the company has since raked in huge amounts of money to the sweet tune of US $70 million in a flotation in both Hong Kong and New York, thus showing once again how eager companies are to line up and break into the Chinese Internet market. All feature web-based email, chat rooms, news – the usual sort of stuff. All the sites are in English and Chinese, and as you might expect, the sites based in Hong Kong and Taiwan are a little more lively and risqué than the more stolid China sites. But again, Xinhua are a major shareholder and content provider – they stage-manage a lot of the content very carefully – so no wonder that for all the hype China.com are 39ᵗʰ and falling among mainland portal sites. In May there were stories that Xinhua ended up blocking China.com on the mainland in the sensitive run-up to the anniversary of the June 4 1989 massacre in Beijing's Tiananmen Square, which reportedly also caused a lot of tension between CIC management in Hong Kong and Xinhua officials. But it is a strange deal that's going on there, a deal which looks all the more strange now that US-China relations are suffering and the US Defence industry looks like it has suffered from hi-tech sabotage. Look at China.com's backers – America Online, 24/7 (a New York-based media company), Bay Networks, Bechtel Enterprises, Edelson Technology Partners, Mitsui and Company, New World Infrastructure, and Sun Microsystems. There is a lot of US and Japanese technology there, and as much as they want to make money in mainland China, are they hurting long-term US and Western interests?

Given scenarios like those above, why do we assume the information future is so bright? I'm no Luddite, and besides it is far too late to turn back the Internet tide now anyway. But in our quest for material wealth and in our quasi-faith in technological progress, it is important that we in the West regain our collective capacity for thought and consider the downsides of technology. We had better be able to start asking and answering some tough questions. Watch what happens in China over the next 10 to 20 years – it could be a harbinger for us all.

36

The Revolution Will Be Customised

Jack Schofield

You've decided to watch *Hamlet*, so you call up the play on your personal communicator. OK, you're short of time, so you decide to go for the short version in American English, with your face superimposed on Hamlet's, and Alicia Silverstone's on Ophelia. But maybe you're having second thoughts about watching a tragedy – you've had a hard day at the virtual office – so you select a happy ending. . . .

Well, why not? If you were going to buy a PC or a car, you'd expect it to be customised to your requirements. Why shouldn't you have a customised newspaper, that has only the news, columns and features you want to read? Why shouldn't you enjoy customised versions of novels and plays, including Shakespeare?

There's nothing new about the idea: every version of *Hamlet* has always been customised for an audience. Hamlet has even been played by women (Asta Nielsen, the divine Sarah Bernhardt), which might have surprised Shakespeare, who would have cast Ophelia as a boy. The text has been translated into American, Russian, Japanese etc., and it would not be surprising if there were cockney, scouse or jive-talking 'ebonics' versions. In 1994, *Hamlet* was even performed live, over the Internet, as *Hamnet*, by actors typing in their lines – a performance that brought tears to my eyes.

```
 * SCENE THE QUEENS CLOSET [43]
<hamlet> Ma: what the fuck's  going on?  [44]
*** Mode change "+v _QUEEN" on channel #hamnet by aurra
<_QUEEN> Don't flame me, i'm your Ma!  [45]
<_QUEEN> Er.... [46]
<_Prompter> Psst!  Thou hast thy father much offended  [47]
<_QUEEN> Oh, right... Yr dad's pissed at u [48]
*** SCENE is now known as Polonius
 * hamlet slashes at the arras  [49]
<Polonius> Arrrrrrrrrrgh!!!!!!!!!!!!!!
*** _Prompter is now known as _Drums
<Polonius> [50]
```

```
*** polonius has been kicked off channel #hamnet by hamlet
    (hamlet)
*** hamlet has been kicked off channel #hamnet by Duck9
    (Duck9)
<_Producer> [51]
<_QUEEN> Now look what u've done u little nerd :-(  [52]
```

 A purist might argue that the 'real' *Hamlet* would have to be an authentic reconstruction of Shakespeare's text, with original spelling, played by an all-male cast somewhere like The Globe, which is a reconstruction of Shakespeare's theatre on London's South Bank. (Presumably the actors would use original pronunciations, too.) But said purist would also, I hope, agree that Grigori Kozyntsev's Russian film, *Gamlet,* is one of the great *Hamlet*s of all time, even though it's not performed on a stage and doesn't use Shakespeare's language. The Japanese anime comic-book version is also still *Hamlet*, and after that, where do you draw the line?

Shakespeare is, of course, an extreme case, but in the future, I expect 'mass customisation' to become the norm for information as well as for consumer goods and services. That's almost inevitable because tomorrow's information, being stored in electronic form, will be almost infinitely malleable. In fact, since there is no view without a point of view, much information may only exist in the form which it is created for and perceived by individuals.

The customisation of information products already exists at a primitive level. For example, you can buy a child a birthday story that includes their name rather than one chosen by the author: a computer simply fills in the blanks, then the book is printed and bound. Today's 'print on demand' systems could already enable novels and reference works to be produced in editions of one, with variations to suit the reader.

Fly with the right airline and you can now make a personal choice of inflight movie, and some will custom-make an audio CD with your personal selection of music for the flight. (In the far east, record shops have been offering similar tapes for decades.)

Many if not most information products are already customised for local audiences, and a Hollywood film may be cut differently to match the requirements of the British Board of Film Classification, video sell-through, and television broadcasting. There's no reason why a DVD disc of a film shouldn't let viewers choose the level of censorship (or sex or violence) they want, the language, and the running time. Directors could even supply all the main footage, so you could create your own 'director's cut' to compete with theirs.

The Internet allows the customisation of information to be taken even further. At the moment this is most evident at 'portals' or Internet gateway sites such as Yahoo! and the Microsoft Network (MSN), where users are invited to create their own personalised pages. When you hit the site, it produces a page with your selection of news topics (computer news and sport, for example), TV schedules and a weather report for the area where you live, and the latest prices for any stocks and shares you may own.

It's important to understand that this page doesn't exist until you look at it, and most probably, it will never exist again. It's constructed 'on the fly' from a range of 'live' feeds and other information that is being updated on a more or less continuous basis. My Yahoo! is a working example of the mass customisation of information in action. A lot of people don't get the Internet, even though they think they do. They think it's an electronic publishing system, so they try to produce the equivalent of printed pages, with carefully spaced headlines, fixed illustrations, and texts in unusual type-faces. Instead they should be generating pages to match the equipment people are using, their interests, and, perhaps, their purchasing power.

Even if people go to web sites using PCs running Microsoft Windows – and at the moment, almost all of them do – they could be using different browsers with different display capabilities, and they may have a different screen resolution set, or different colours. They may have chosen to display text smaller or larger than normal, which will 'break' a lot of page layouts. They may have Java or JavaScript turned on or off. They may or may not have installed browser accessories or 'plug-ins' such as Macromedia's Shockwave Flash and Real Networks' G2 media player. They may be using a high-speed line – with ADSL or a cable modem – in which case they may want 3-D graphics and moving video, or they may be dialling in with a slow modem from a hotel room in Siberia, in which case they'll probably want plain text, and as little of it as possible.

Web sites need to detect as many of these variations as possible and create pages that make the best possible use of the display capabilities available. Either way, the attempt to serve the same web page to people with, say, 3Com Palm handhelds and webTV sets and Unix workstations with huge colour screens is insane.

In the future, the situation is going to become even more difficult for publishers fixated on print media ways of doing things. Over the next five years, analysts expect a huge growth in the number of Internet-enabled TV sets, handheld computers, mobile phones and a variety of WIDs (wireless information devices) that don't work like Windows.

Customising content is a little trickier, but should become more practical in the future. Web sites can store 'cookies' (small pieces of text) on the user's computer to record a log-in ID, a list of the pages viewed on earlier visits, and other significant information. For example, if someone visits a mall to buy computer supplies, they can be served an opening screen that features computer supplies, whereas if they buy clothes or flowers, they could get those instead.

Some sites already use extra information to provide customisation. For example, if you use the Air Miles UK site to check how many air miles you have accumulated, the site serves up special offers that might appeal to you. People with 500 air miles to spend get different pages from people with 5,000.

The problem, of course, is that most sites don't know very much about their visitors. That could change as web site providers compile and correlate information from different sources – which they are inevitably going to do – or as web surfers start to use profiles or turn to 'intelligent agents' to represent them in cyberspace.

Agents are 'software robots' that act (or appear to act) independently. The most common example is the Microsoft Agent usually represented by an animated paperclip in Microsoft's Office suite; that and different animations of the same agent are also used on more than a hundred web sites. Other 'bots' are used, for example, to collect and compare prices from different web sites, to help shoppers, and in games like Quake II. People will, I think, start to use agents because otherwise they won't be able to cope with the web: it's already too big to be manageable, and it's going to get a lot bigger.

The size of the web accounts for the fact that a lot of the Internet's earlier 'truisms' – that all companies were equal in cyberspace, that intermediaries like travel agents would become dispensable, that electronic commerce would take place in a 'frictionless' economy, and so on – have so far turned out to be false. The fact is that anyone who wants to buy a book is most likely to go to Amazon.com, and a couple of dozen well-known and heavily-branded web sites have a lion's share of the traffic. . . .

Why should this be so?

First, you have to go beyond Moore's Law (which describes the decreasing cost of computing, as chips became more powerful) and Metcalfe's Law (which explains the increasing usefulness of networks, as they become larger) and start reading the Nobel Prize-winning economist Ronald Coase. In his article 'The Nature of the Firm', Coase explains that the way firms operate, and ultimately their size, depends on their transaction costs. Companies work because they can do some transactions (perform some actions) cheaper than individuals, and they stop growing when transactions can be performed just as cheaply on the open market. In other words, you go to a travel agent for a plane ticket because it's cheaper (in time and / or money) than phoning all the different airlines and trying to find the best price for a particular trip. If anyone can get the best price in seconds from a web site, the travel agent's business gets smaller and perhaps ultimately ceases to be.

In theory, the web provides individuals with access to as much information as most companies, and transaction costs that approach zero. The problem is that very few web users are capable of finding the information they need, and finding the right information may be more trouble than it's worth. If you get 600,000 'hits' from a search engine like AltaVista, the transaction cost of finding, verifying and using the appropriate information is only zero if you have a free phone line and nothing better to do for several hours . . . or maybe days.

When the web becomes 10 times or a thousand times or a million times bigger than it is today, even the most skilled searchers will decide there's too much information to handle manually, and they'll use intelligent agents to search and shop on their behalf. And in order to function efficiently, these agents will have to be your representatives in cyberspace. They'll have to know what you like, or what you might like, and they'll be willing to trade this information with agents representing suppliers and other web sites.

Agents will also learn from one another, using techniques like 'collaborative filtering'. This technique was pioneered by sites like Ringo at MIT, which recommended music, and is used today by shopping sites like Amazon. At its simplest, collaborative filtering is based on the idea that the sort of people who like one sort of thing will tend to like other sorts of things too. For example, if you've been buying CDs by John Lee Hooker, Muddy Waters and Bo Diddley, there's a good chance you'll like Albert King and Etta James, even if you've never heard of them. Buy three rap albums and you could be offered a special deal on a back-to-front baseball cap (no extra charge) and a pair of Air Jordans.

Agents will solve the problem of mass customisation by providing information suppliers with all they need to be able to send you appropriate information in an appropriate format. You can certainly expect some sites to tailor their content, too, as newspapers do today.

Would you like a Tory, Labour or Marxist-Leninist (or Democrat or Republican) take on the news? Did Manchester United deservedly crush the has-beens from Arsenal, or were the Gunners robbed by a half-blind referee? See your local paper for details. . . .

If suppliers don't customise information, users will.

Your agent could specify, for example, a language and reading age for text, and when agents become more powerful, they will translate it as well (or get translation 'bots to do it for them). Today, for example, an agent looking something up in Microsoft's Encarta encyclopaedia could choose the most appropriate of the nine editions – the American and British ones, for example, have different approaches to baseball and cricket – but there's no reason why an agent couldn't extract and summarise an article, then translate it into Swedish, or whatever.

I'm not claiming all this is going to be easy – translating a sentence like 'Wait at the Pelican crossing until you see the green man flashing' is always going to require a lot of embedded knowledge – but the results don't have to be perfect. They only have to be better than the alternative, which may well be nothing.

Ultimately, however, I suspect we'll be faced with the problem with which we started – Which is the real *Hamlet*? – only worse. The good thing about *Hamlet*, from this point of view, is that the text can be more or less fixed to a specific author, time and place. Given access to a time machine and a camera, you might even be able to grab a definitive version. In the future, it will often be impossible to fix information in any such state. Web data will be updated at different rates at different times, then assembled into more or less unique packages for delivery to individual consumers . . . who will then modify and display it according to their own needs and desires. Give it an ISBN and Dewey decimal your way out of that, if you can.

The prospect of a Daily Me newspaper has already worried some commentators, and the idea of My Personal Internet may horrify them. However, people today are free to read only what they want to read, watch only what they want to watch, so it may not make much difference. Anyway, most people want to learn new things – they will pay a premium for novelty – and the web will still offer more variety than any previous medium.

Some people are also worried about the problem of privacy, and this is a valid concern, but agents need not necessarily make the problem worse. (What privacy do you have if your spending can be tracked via credit and store cards, and your movements followed using CCTV cameras and the signals sent out by your mobile phone?) Agents can be anonymised, and they only need to carry credentials, not an identity. Tesco doesn't really care who you are, only how often you buy beer or toothpaste, and whether or not you have the funds to cover the bills.

37

Information Wars – F is for Fake

Robin Hunt

Introduction

Oh bloody hell, the future of information. Again. Another opportunity to be a guru or a fool – or both. As others, historians of information science, have written in this book, visions of the future aren't so much bleak or fantastic, just inevitably wrong.

In *The End of History and the Last Man*,[1] Francis Fukuyama's ground-breaking bestseller of 1991, he posited the end of ideological conflicts, the triumph of liberal democracy. Remember what followed? The Gulf War, the Bosnian tragedies of the same period, the struggles in Burma, Ireland and Beirut to name just a few. Remember those ideological conflicts – they were once news, *information to be captured,* to be analysed by people such as Fukuyama. Remember the certainty felt about those wars, the intellectual and moral positions taken?

Remember, perhaps, more recently, the plea – as this author heard being chanted outside the Embankment tube station in the spring of 1999 by a group of a 100 or so Socialist Worker Party members – to 'Stop the bombing [of Kosovo], support Milosovich'? Scary stuff, certainly, for non-believers of the great US/Nato conspiracy in the Balkans (itself, another 'fact', more *information*). These are remembrances we all have; news we have consumed in one way or another; histories distorted by time; facts as yet not disproven, or already ridiculed – but undoubtedly stored somewhere *digitally*, be it in a newspaper's library or a police surveillance unit's video storeroom; undoubtedly stored somewhere *mentally*, be it as a real experience or as seen on CNN or Beirut television.

Well, this panorama is the certainty for 'the future of information': in an era of plentiful information, an era of plentiful positions – ideologies, histories, views of the world, *beliefs, conspiracies,* the tabloid press and Public Service television, repressive governments and anarchist web publishers – information, however it is distributed, collected, sorted and maintained, is at war with itself.

In this war the troops aren't kevlar-suited, night-envisioned commandos, nor ragged Uzi-toting fundamentalists. In this war there aren't even commanders, not tacitly so anyway. In this war facts and editorialisations, the focus group and the finger in the air, graphs ripe with e-commerce consumption patterns and endless high-street 'Sale On' signs and posters, satellites and modems circling and proliferating respectively, proprietary technologies and the 'open source' ones: these are the weapons.

And information is the battleground, and control of information the ultimate objective. The ultimate objective when for many the world is now a place in which we know we cannot *know enough*. Where we know we are dumbed down because we are so aware of the immense wealth of information which we haven't time to consume. Where we make decisions about *knowing*, about *what we choose to know*: Manchester United or the launch of the Euro, both but nothing about politics; share prices but not unemployment figures. We have to make these choices – there is no time to be a know-all; there never was, but now we know this fact – its signifier the magazine rack in any newsagent, the programme schedules for satellite television, any decent bookshop, the Internet. The signified being our deep-seated paranoia about information, our turning towards so much that is non-empirical: from solar eclipses to millennial cults, new or ancient religions, through yoga, the lottery, and mass drug consumption as a weekend activity. We are bored, we are excitable. Most of all we are messed up, dislocated.

Further, this is a world in which information as formal as $e=mc^2$ is now as relative as quantum physics or Julia Roberts' boyfriends. This is a war of high culture and low, news and entertainment, share prices and share price rumours, new projects for third-millennium social order and the endless diaspora of refugees across Eastern Europe, policy units and pragmatic entrepreneurs, sports rights and open government, *The Times* and the bulletin board.

This is a Darwinian fight in which the million year dynasties have been replaced by the millisecond. In which the gene has been replaced by the meme[2]. In which Bill Gates' vision is forever made foolish by Third World debt. In which his salary could solve Third world debt. Where the *vision thing* is as useless as the astrologer; the genetic engineer as useful as the plumber. A place in which we all want to find information. And where truth is useful too.

So this article isn't about smart houses, replicants, supercomputers or the death of the Library, the book, the written word. Not about Windows 2001 nor AOL Mars. Nor even the wonders of the Internet. Instead it is about lies.

1. Secrets and Lies

1. Breakfast time in Notting Hill. A familiar floppy-haired man opens his front door in his boxer shorts, feeling pleased with himself: he has, after all, just slept with the most famous film actress in the world. And he just a humble bookshop owner. He opens the door to be greeted by two hundred paparazzi. He has been *shopped*. His louche flatmate has been boasting down the pub.

Don't worry, Hugh Grant tells the incandescent Julia Roberts, today's newspaper is tomorrow's fish and chip wrapper. She is not convinced. Every interview, every *Vanity Fair* or *Details* writer or David Letterman show researcher will now have at its disposal these photographs, this story, she says. It is news, it is information; it is databases and Internet sites. It is fact. Part of the story. There is no such thing as short term memory now. Everything is captured. Forever. On a computer.

2. Lunch time in Kensington. A familiar plumb-filled voice declaims in the *Evening Standard*. Brian Sewell, the paper's art critic is angry about a profile written of him in the famously high-brow *Independent* newspaper, by their Feature Writer of the Year, Deborah Ross.[3]

'Her first questions were aggressive enquiries into parts of my life that seem to me my business and mine alone, prurient and pushy – and so it continued, as though she were writing for *Hello!* rather than the respected Indy, one question about work, three about sex, one question about dogs, four about sex… 'Do you cook?', followed by 'Are you gay, do you sleep with women, when did you first masturbate, have you any fetishes, are you into bondage or do you have other kinks?'[4]

Not, perhaps, the kind of interview that a man famous for his love of the Four Last Songs of Richard Strauss, Caravaggio's Flight into Egypt, might approve of, or enjoy. Nevertheless these things happen to the famous – and the not so. But then, there's this:

'Had all this been done with the practised skill of an assassin with a bodkin, her prose as polished and damascened as its steel, her performance exquisite, I could have forgiven her patronising spite, even admired or envied it, *but she introduced so many errors of fact, culled so much nonsense from other sources* and wrote so badly that forgiveness is the last thing in my mind.'[5]

3. Breakfast time in Washington, the White House, sometime a long time ago. A familiar flurried, florid man is checking his hair. Next door his retinue of spin-doctors are checking the web (again): 'As required by Sec-

tion 595(c) of Title 28 of the United States Code, the Office of the Independent Counsel ('OIC' or 'Office') hereby submits substantial and credible information that President William Jefferson Clinton committed acts that may constitute grounds for an impeachment.

The information reveals that President Clinton:

- lied under oath at a civil deposition while he was a defendant in a sexual harassment lawsuit;

- lied under oath to a grand jury;

- attempted to influence the testimony of a potential witness who had direct knowledge of facts that would reveal the falsity of his deposition testimony;

- attempted to obstruct justice by facilitating a witness's plan to refuse to comply with a subpoena;

- attempted to obstruct justice by encouraging a witness to file an affidavit that the President knew would be false, and then by making use of that false affidavit at his own deposition;

- lied to potential grand jury witnesses, knowing that they would repeat those lies before the grand jury; and

- engaged in a pattern of conduct that was inconsistent with his constitutional duty to faithfully execute the laws.

The evidence shows that these acts, and others, were part of a pattern that began as an effort to prevent the disclosure of information about the President's relationship with a former White House intern and employee, Monica S. Lewinsky, and continued as an effort to prevent the information from being disclosed in an ongoing criminal investigation.'[6]

And we don't need to go much further to know about facts, disinformation, rumours, and statements on oath and legal depositions to know where this one went. Let's just remind ourselves of Matthew Drudge and how his web site, the Drudge Report[7] ran many, many stories about the President's friend. The dress. The illegitimate child. The definite impeachment. The escape. Hilary for President. Some were right, some wrong. Nevertheless, the 'Zippergate' scandal and the role played by new types of information providers, brought this response from the *New York Times* last Christmas (1998):

'There was Matt Drudge, who first posted the story, pilfered from *Newsweek*, on his Internet tip sheet; the talk shows that gave it legs; the 24 hour cable news networks that rode it up the ratings, and the handful of respected newspapers that publicly stumbled in the scramble to keep up.

Looking back on the past 11 months, many people inside and outside journalism described the episode as a laboratory experiment for the post-modern media age – the moment when the diverse forces of the instant communication era converged for the first time on a major political scandal.

How members of the media have performed – and how the public fared – is a subject of stark disagreement. After months of criticism, many journalists now express a sense of almost vindication. Most of what was reported, they said in recent interviews, has turned out to be true.

Others, however, described the last 11 months as a "very dark chapter in the history of the American press," as Geneva Overholser, a former ombudsman for the *Washington Post*, put it. After such shameful excesses, some said, it is an inadequate defence to point out that the reports were not wrong after all.'[8]

These new news sources aren't going away: technology for cable or the internet is cheap. There are audiences for this stuff. What was once, say, the provenance of the alternative press, the 'counter-culture' , is now the mainstream: the 'shop-counter culture'. Think of the *National Enquirer* magazine – it reported in June that Princess Diana's spirit was present at the wedding of Prince Edward and Sophie Rhys-Jones. In fact the story was its front page splash. Does anyone believe this? Who knows? Does anyone buy this? *Millions of people.*

However, the most telling piece of information about the whole Clinton affair was far simpler than technology outlets or ghostly apparitions down Windsor way: the most powerful man in the world said things – used language – on oath which was not true. Yet, when he said he did not have sex with Monica Lewinsky, he was in a sense, a legal, twisted black-means-white sense, telling the truth. Everything in language is, as linguists have been stating for hundred of years, about the relationship of word and meaning. Collective meaning, cultural meaning, legal meaning and – heaven forbid – ironic meaning. It is about a search for 'truth'. A truth which is relative now and increasing numbers of people realise this 'fact'.

A simple question, for example. Who invented the telephone? And the source to verify this question we use is the multimedia encyclopaedia, Encarta, from Microsoft corporation. First we try the British, American and German editions and the answer is Alexander Graham Bell. But look at the Italian version and the credit goes to Antonio Meucci, who allegedly beat Bell by five years. Meucci was an Italian, an impoverished candle maker. As *The Wall Street Journal* reported: 'Technology [such as the Internet] and globalisation are colliding head-on with another powerful force: history.'[9]

Whether the issue is the light bulb, who discovered the AIDS virus, the ownership of an island on Korea's southern shore, who won the battle of Waterloo, or the 'existence' of Kurdistan, Encarta has differing answers depending on the local market, the local *thinking*. The local thinking *right now*. Technology now has to deal with the issue which has bedevilled literary and social theory (to name but two academic schools) objectivity. But unlike the ivory-towered academic who can build a career around the existence or not of a chair, technology is about the free market and sales. What global information provider wants to stir controversies in local markets?

Reporting on this issue the *Wall Street Journal* closed with a baffling statement: '… the universality of the web also frustrates efforts to localise content. And there remains the possibility that it will bring about pressure for one, *universally applicable version of history*.'[10]

Unless the world becomes a very strange homogenous place that is simply nonsense. History, like its broader church information, is at war, continuous Maoist revolution: nothing can ever be universal again. Nothing ever was.

4. May 1999, a room in the *New Republic* magazine office sees the sacking of an associate editor, Stephen Glass, for making up a story about teenage computer hackers being in such demand that they have started hiring agents. According to the magazine there are 'potential problems' with some of the 40 other articles Glass has written for *New Republic*. His work has also appeared in publications such as *The Washington Post*, *Harper's* and *GQ*.

5. The jury is decided, Carlton television's documentary *The Connection*, about a supposedly new drug route into Britain from Colombia will win the award. A little later a *Guardian* newspaper investigation presents much evidence that the programme is a fake. The details, many of which are disputed, are now a matter for Carlton's internal investigation. Some of the key points can quite easily be checked. Did the programme-makers pay for the airline ticket of the so-called drug 'mule'? Did he fail to get into Britain, and was the man portrayed as the number three in the Cali drugs cartel really a retired bank clerk?

More worryingly, perhaps, how many of us care?

Bloody hell, this information future could look very strange indeed.

2. The Girl in the Picture

This warring information is not just about text, histories and databases and newspapers, it is pictures too: 'One of the official photographs from the wedding of the Earl and Countess Of Wessex was digitally altered. On the instructions of Prince Edward, Sir Geoffrey Shakerley replaced the frowning features of Prince William with an image of him smiling, taken from a different photograph.

- When Wilhelm I of Prussia visited a Berlin livestock show in 1861, he wanted a photograph of himself with the farmers. There was no photographer present, so the 30 farmers were later reassembled and photographed, and a shot of the king was pasted in.

- Trotsky was manipulated out from countless photographs on the orders of Stalin. One picture was retouched so hastily that Trotsky's elbow hovers next to Lenin.

- Mussolini was terrified of horses, but felt that a picture of him on a stallion would be good for his image. A picture was taken and a groom holding the reins of the horse was airbrushed out.

- After the Iran-Contra scandal, the White House press office airbrushed Oliver North from a photograph taken with President Reagan in the Oval Office.

- In April 1996, John Prescott was outraged by a doctored photograph published in the London *Evening Standard* of him and his wife at a party. He said: 'The bottle in front of my wife has been cropped to make it look like a champagne bottle.'

- In 1997, the *Daily Mirror* published a photograph which appeared to show Diana, Princess of Wales and Dodi Fayed in an embrace. In fact, Fayed's head had been rotated, using computer technology, to give the appearance of intimacy.

- Last year the cigar was removed from a picture of Isambard Kingdom Brunel used to publicise the restoration of SS Great Britain. The decision was taken because smoking was considered 'politically incorrect'.

What in fact is the case with photography is that manipulation of the image began almost with the birth of photography, yet for decades its cultural significance has been as a mirror to the world, a reflection – *verité* – rather than its real function as a meditation, a snap shot of something which is in perpetual motion, a fake if you will. So while photography begat abstraction as an artistic principle, just as much as the motor car begat the Futurists, or Freud the photomontagers, its true achievement has been to make us as Coleridge wrote, suspend our disbelief. The curator of the Bradford Museum of Photography, Russell Roberts, put this well recently: 'Photography has been the main agent in the fabrication of history, it shows us a very limited view of the past. There have always been editors, curators and individuals who decide what makes it to the front page.'[12]

Nowadays everybody knows that most of the extras on the movie *Titanic* didn't exist outside a microchip and that without the latest computer graphics the film could hardly have been made, and still we cry (allegedly). With fiction, the onward march of special effects presents no problem. It is when the 'special effects' – the recreations – invade the real world of journalism that the worries start. Kenny Lennox, picture editor at *The Sun* newspaper, told the *Independent* in 1997[13] that he had been offered a shot of O. J. Simpson praying at the graveside of his murdered wife – awkwardly it had a person in the foreground obscuring Simpson's head. Another magazine offered him a perfect image of the visit. But it had been created on a computer, using the top of O. J's head from another photograph. What to do? Well, for now picture editors tend to say it is bad practice, but the examples above suggest that not all is strictly accurate in the media world of images.

On the Internet it is possible on most days to find images of naked celebrities – sometimes images from movies or *Playboy* spreads, or paparazzi beach shots; but very often the personality's face is 'Photoshopped' onto some *Playboy* pet's body. So prevalent are these 'fakes' that some sites *just* feature them, with ratings for the best and worst ones.

In December 1994 the author was part of a project at *The Guardian*'s Product Development Unit in London. We created a 'newspaper of the future', *G2004*. There were four versions, each personalised to the profile of mythical readers: a student, a home worker, a trade unionist etc. As a joke we made up listings for a series of personalised television programmes: Arnold Schwarzenegger playing for Arsenal; or versions of old football matches with different results – that way Manchester United could win every match, or not; the Spice Girls could be naked in their videos or not; William Hague could even be Prime Minister in this world.

The response was mixed, most hated the paper's hermetic world. Internally, the message that information, journalistic truth, could be manipulated for a Panglossian best of all worlds approach to readership, was not well received. But it was probably taken on board. A little. Recently the editor of *The Guardian* wrote about trust: 'The digital age is truly exciting. I do believe that the Internet is a revolution in communications to rival Caxton and Baird. But, ultimately, when stripped of all the silicon hype, we are just talking about the communication of information. And, ultimately, all that matters is the simple question: is it true? Can I trust this source? If you can be trusted you will win in the end. If you cannot you will be no better than any other cracked voice shouting for attention. When you walk through the gateway of your sister school of journalism in Colombia, you pass under an inscription by Joseph Pulitzer that reads, 'a cynical, mercenary demagogic press will produce in time a people as base as itself'. There has always seemed to me to be little point in being a journalist – or in wanting to be a journalist – unless you believe that the opposite is also true.'[14]

Now that isn't the future. The future of information is that we as users will grow ever wary, ever more untrusting of it – we are its consumers, and like all consumers these days we demand the cheapest, the fastest, the best. Sometimes we will want diverting information, such as *Hello!* magazine; other times accurate information about our pension fund. Sometimes we will want to understand the Balkans, other times to imagine they don't exist. And sure we do trust our media, information choices, our preferences – but only as far as we can spit. Increasingly, we take a pinch of salt with everything we are told.

Does that make us cynics, or under-informed, or dumb? Probably, but these phrases are certainly also 'true' for most information providers in the twenty-first century who will be busy creating histories – 'Do you remember when people believed us, our information, our world view, our photographs, our writers, way back in 1985?' These histories (and perhaps we will see them in the British version of Encarta 2010) will be all about an accepting, passive – dare one say mass – audience for information.

Which, of course, never existed.

References

1. From *Kirkus Reviews*, November 15, 1991. In 1989, *The National Interest* published 'The End of History?' by Fukuyama, then a senior official at the State Department. In that comparatively short but extremely controversial article, Fukuyama speculated that liberal democracy may constitute the 'end point of mankind's ideological evolution' and hence the 'final form of human government.'

2. *The Selfish Gene*, revised edition, 1990. In Chapter 11, 'Memes: the new replicators', Dawkin's ideas diverge from the central physical replication discussion, classical Darwin and enter into the area of the replication of culture and ideas; about how ideas have their own gene pool just as potent as DNA.

3. '"Who is Deborah Ross?" I asked – to which came the answer that she is a sparkling, witty, brilliant, accomplished, penetrating, waggish, racy and sympathetic writer of features, and indeed has just won the Feature Writer of the Year Award. How, I wondered, could she be all these and yet be quite unknown to me, though I read the Indy every day of the week. How could I have turned the page so swiftly? How could I not have been held in weekly thrall by this direct descendant of Addison and Steele? I now know the answer, for with the flattery that followed and the considerable pressure of a friend in the employ of her eximious newspaper, I succumbed and consented to be interviewed by this jewel among journalists.' Brian Sewell, 'She Mocked Everything I Said.' *Evening Standard*, London. May 5

4. ibid.

5. ibid.

6. From, because the Excite search engine found this one first, http://CNN.com/starr.report/

7. http://www.drudgereport.com

8. 'Scandal Coverage, Risky Era for New Business,' by Janny Scott.*New York Times*, December 24 1998, pp A16

9. *Wall Street Journal*, Friday June 25, 1999, pp. 1

10. ibid

11. 'Bush telegraph on . . .: doctored photographs.' *The Daily Telegraph*, June 99

12. 'Exposed: the Camera's White Lies.' *The Sunday Times*, June 27, 1999, pp. 14.

13. 'The lady vanishes … : In the age of computer enhancement, can we still rely on the news photograph as documentary truth?' by Emma Daly. *The Independent*, Media section, 7 July 1997.

14. Alan Rusbridger, 'Who can you trust?' *Aslib Proceedings* 51 (2), February 1999, pp. 37-45

38

Musings On the Future of Information

C. David Seuss

All Human Knowledge in a Single Integrated Search

From time to time while on a Star Trek mission, Captain Picard or Captain Janeway will speak into the air, address the ever-present Computer, and ask it a question about science, history, or culture. The Computer always has the requested information, and never responds, 'Which database would you like me to search?' The Computer in Star Trek has all human knowledge at its disposal.

And why should it not? Admittedly, human beings are creating more and more information faster and faster. But meanwhile, computers are getting more powerful and more connected at an astonishing rate. If computers get powerful enough and connected enough, then sooner or later it will be possible to load everything everyone knows on them. It's a race between, on the one hand, the creative labors of the human race and, on the other hand, the technical prowess of the designers of microprocessors and the attractiveness of networks.

I for one am betting on the networked computers.

One indicator of who is winning the race is the cost of computing. The cost of a CPU cycle has fallen to one ten thousandth of what it was ten years ago. There is no reason to believe that it will not fall another ten thousand times over the next decade. In one generation, the cost of the most basic process facilitating human commercial and intellectual activity will have decreased by *100 million times*.

Also, consider the rate at which computers are being networked. Obviously, the more networked, the less important it is that a single computer has all the information being sought. Before 1994, the number of computers on a large network was measured in the dozens. Then the World Wide Web formed out of the void. Suddenly, there was network with tens of thousands of computers on it, then hundreds of thousands, then millions, and now there are more than *100 million computers all on one network*.

This is a profound change in the capacity of the human race to store and share information.

It will not be long before it is technically feasible to put all human knowledge on a computer network accessible to everyone on the planet. Frankly, I think it already is. Information technology and its adoption by the world has achieved the capacity to do what the Star Trek Computer does three centuries from now. Now if the physicists can just get to work on the transporter and faster than light travel!

User-Generated Content

Publishing content has always been a costly undertaking. In 1455, the cost of the Gutenberg Serial Number One printing press was 800 gold florins, which equates to $28,000 of gold using 1999 prices. By the year 1500, there were 250 printing presses in Europe and over 8 million books had been printed, more books than had been printed in all of history before 1455. But the presses and publishing processes were expensive, and the owners of these presses could only afford to publish content with really big markets, like the Bible. This established the basic structure of the content industry, publishers would create content that could be sold in a standard format to large numbers of content users.

Hundreds of years passed without any real new idea for publishing content. The presses got faster and the cost of a unit of published content went down, but the structure of long-run, broadly-targeted content manufactured by a publisher stayed unchanged.

Then in 1985 or so, the personal computer industry started to change things. A personal computer, a desktop publishing program, and an inexpensive Hewlett Packard printer became the minimum efficient unit of scale for a publishing operation and a newsletter industry was born overnight. Now the number of users interested in a piece of content could be much smaller for the content to be published. But one thing did not change, that being the idea that content should be created by a publisher and delivered to a user group. This was because the labor to assemble and disseminate the content was high relative to the amount of content any one user might have an ability to contribute.

The web creates the possibility for a fundamental shift in the paradigm. It has always been true that users of content have vast amounts of information about the subject in question. The purchasers of cookbooks have their own recipes, job searchers know a lot about the companies they have worked for that would be of interest to other job searchers, readers of software reviews are much experienced with many software programs, users of industrial product catalogues know a great deal about the strengths and weaknesses of different suppliers.

With the web, virtual communities, and email, the work required to author, assemble, and disseminate content is inconsequential. Suddenly, it is possible for users of any specific content to band together and share their great depth of knowledge. Databases of user generated content are starting to appear; in a few years there will be tens of thousands of them.

This changes what it means to be a publisher. Publishers in the future will be creators of content for users; they will be aggregators of user-generated content. This is a truly new idea, maybe the first new idea in the content industry since Gutenberg started printing his bibles.

Multimedia Data Types

From 1455, when Gutenberg invented the first printing press, until around 1950, the printed word was the dominant form of information transfer. Then around 1950, the widespread commercialisation of TV began pushing the printed word aside.

By 1999, it is an MTV, video-bite world – unless you are in the traditional online information industry. Somehow, we are still in the age of the printed word. Search an archive, and you get articles, not video clips. Obviously, this has to change, or the online information industry will be increasingly out of step with the world. In case you have not noticed, the web is very much a multimedia experience.

There are several reasons why the online information industry is lagging behind the world in the arena of multimedia data types. Partly it is storage cost. A disk farm with enough storage to hold even one year of news clips would run into the millions of dollars. Partly, it is transmission bandwidth. It could easily take hours to download a selection of interesting video on a low-bandwidth modem. Partly it is search technology that is text-oriented and the fact that reliable speech to text converters have not been available to us to populate a database index.

But all of these factors are changing. Storage costs are dropping by orders of magnitude every few years. The appetite for bandwidth on the part of the public is inexhaustible and the telecommunications industry is organizing to supply it. The wide spread adoption of close captioning for the hearing impaired provides a searchable text stream for indexing purposes, at least in the news arena.

The first indicator that the future has arrived will be the deployment of an online, on-demand movie library (entertainment applications seem to always drive new technology in the PC industry). When this happens, know that the future has arrived and the online information industry had better get on board.

Research Skills Become Paramount for Everyone

It used to be that library research was something that most of us did little of during our school years (maybe once a year for that term paper), and almost never in our personal and work lives. Most people left library research to the professionals who did it all day long.

But the web, the mother of all online libraries, has changed all of that.

More and more of us have become 'wired'. We have integrated email and the web into every aspect of our business and personal lives. We check email a dozen times a day, feel a let-down when there is no new email, and immediately check it again. We use the web to research clients and competitors, to find vacation lodging, to learn about our medical conditions, to contact potential suppliers of services our businesses need, and to buy personal items. We move seamlessly from business decisions to personal decisions, from working at the office to working at home to working from a hotel, with the only constant being that we go to the web almost every time we need to know something.

But this reliance on the web does not come easily. The absolute first impression that the web makes on new users is that there is a boundless lode of information on any topic just for the asking. At first, new users bounce from hyperlink to hyperlink, feasting on data. But before long the reality starts to set in. Finding information in an electronic library can be hard, especially for the unskilled. There is too much information, the most relevant documents are awash in a sea of not so relevant, poor quality documents. The new users mature and start working on search skills. Sadder but wiser, now searchers not surfers. The important search skills include the ability to frame queries, scan results list, and identify high quality information sources. Once the exclusive domain of professional librarians, everyone has to develop these skills now.

There have been earlier shifts brought on by technology in the skills required of individuals. Twenty years, to get along as an executive in the modern world, all a person had to be able to do was speak persuasively. Secretaries did all the typing and written documents would be iterated many times for review before they were ready to be published. Then the personal computer came along. Writing, and surprisingly, typing became critical skills for executive success. Also, we all had to become our own technical support professionals, working directly with the technology to solve the seemingly never-ending succession of hardware and software problems on our personal computers. Our workdays became dominated by written correspondence that goes directly from our fingertips to the world with the last mouse click on that 'send' button.

Now in the wired world of the web, researching has to be added to the list of essential competencies for everyone. If a professional or manager cannot use the web to efficiently access information, he or she will never be well informed and will not be making the best decisions for his or her business.

My nephews are six and four. The skills I want them to have when they grow up would be those of a smooth-talking, fast-typing, hacker research librarian. Then their success in the web-based world will be assured.

Everyone Knows Everything About You

There is a lot of concern about the loss of personal privacy as a result of the web. What most people are worrying about is that commercial enterprises will be able to track the behavior of individuals, usually for the purpose of targeting some sales pitch. Also, people worry that governments will be able to keep us under some sort of surveillance by tracking our every keystroke.

These concerns are not without some validity. Microsoft once had a plan, scrapped when there was a public outcry (or so they say), to collect information without our knowledge from all of our hard drives about the software and peripherals we have installed. Certainly one believes that both Clinton and Starr would have used every means at their disposal to monitor the other.

But I think everyone is missing the real loss of privacy that comes with the web, and that is the complete transparency that our lives will have to the casual acquaintance, the merely curious, and to the entertainment-seeking surfer.

My company, Northern Light, currently archives several thousand news sources and makes them available to anyone with a web connection. These sources include newswires, regional newspapers, news magazines, college newspapers, press releases, and broadcast news transcripts. Other companies will be doing this someday, individually, or as a result of the sum of the efforts of every organisation with a web site indexed by search engines.

What this means is that if you ever do anything that results in a mention in a published news source, everyone who ever meets you will be able to learn about it. For achievements you are proud of, this is great. But be accused of a crime, participate in a scandal, disappoint a constituency, or flub an opportunity and it will follow you around forever. For example, consider the people who will be able to learn all about your past: your potential spouse, your neighbors, the parents of your son's girlfriend, a hiring manager, Aunt Sally, your florist, and the people inviting you over to dinner on Friday night.

The implications for this are far reaching. First of all, there are greater rewards than ever for good behavior. Manners, good works, and clean living may become fashionable again. And there are certainly more penalties for bad behavior. Before the archiving of news stories in electronic databases on the web, one could start over by keeping a low profile, moving to another town, or joining the French Foreign Legion. But these methods of starting fresh will work no more because everyone you meet in your new life, including your fellow Legionaries, will be checking you out on the web.

The Long Boom

The World Wide Web has wrought a revolution in the economics of production by changing the fundamental costs underlying the processing of transactions. The requirement to process a transaction exists in every commercial exchange. Depending on the complexity, the cost to take an order, create manufacturing and shipping instructions, and invoice the customer has ranged from between $25 and $100 in constant dollars probably since the dawn of the industrial age. Most companies have 10% or more of their revenue tied up in the order processing cycle. And most companies have profit margins of only 4% of sales or so. A portion of this 4%, maybe 2 points, gets reinvested in growth opportunities. The 10% being spent to process transactions is a big drain on the funds available to invest in the future.

Enter the web. E-commerce allows a company to divide all of the work between its customers and its computers. Suddenly, a penny is an inadequately gross unit to measure the cost of a transaction. Ten percent of every dollar of sales is suddenly available to stimulate demand with lower prices, increase investment in growth, or be paid to shareholders who can then invest in new companies. The amount of money available in the economy to invest in growth more than quintuples. This is the equivalent of greasing the gears in the macro economic machine. The machine is going to run faster and faster.

There has been one time before in history in which there was a change in the economics of production as fundamental as this: the Industrial Revolution. The factory system of manufacture was introduced in the first decades of the 1800's in Western Europe, and most aggressively in England. The result was a doubling of per capita income in England between 1830 and 1860, outstripping the other nations in the region by wide margins.

We are on the threshold of a greater and faster surge in economic growth than the Industrial Revolution brought about. Because of the web, the next two decades will see a doubling, then tripling of per capita income for everyone on the planet. There will be 20 years of prosperity unequaled in the history of the world.

The Long Boom has begun.

39

Creating Creators

John S. Driscoll

Just down the street from my home in Rye, New Hampshire, within a stone's throw of the beach, stands the Rye Cable House. It's a nondescript gabled structure of that type that is common along the coast of New England. This one, however, made an indelible mark in the history of communications.

It was at this location in 1874 that the first direct transcontinental cable was completed, connecting Britain, Newfoundland, Nova Scotia and the United States with 3,104 miles of what was called 'duplex cable', enabling two-way simultaneous messaging.[1] The cable station operated from this house for 47 years, initially manned by 16 telegraphers who came from England, Scotland, Wales and Ireland. Behind my house is the street where many of them lived, now suitably named Cable Road, its paved surface running right up to the ocean sand.

Surely there were commercial reasons behind this painstaking project. But part of the impetus also derived from an inner desire we have to communicate as rapidly as possible with those near and dear to us as well as those in faraway lands. We either have information to convey or we have a need for information someone else possesses. The exchange enriches the participants.

Throughout history we have seen various advances in the means of expression: drawings on clay tablet, papyrus, cuneiform, alphabets, printing presses, Morse code, wire and wireless transmission; radio, still photography, cinema, video, the computer and the Internet.

With or without wires, we now have nearly instant access to words, sounds and images, thanks to the digital revolution whose full impact may not be realised for decades. We don't know fully how this new plateau of information flow will impact our lives and our world.

What we do know is that the concept of instant interactivity first manifested in a small way in the transcontinental cable and later popularized by the telephone is now so engrained in the communication process that it has changed the dynamic. No longer is the average person just a receptor of information.

The consumer is becoming a creator. Power truly has shifted to the people.

In the near future we will see an emergence of expression from grassroots levels that will enhance democratic processes and challenge those that are not.

Average citizens today have tools that never were at their disposal. In effect they have their own printing presses in the form of computers connected to the Internet. They have the means to tell their stories to large numbers of people near and far – and that ability changes the traditional relationship between consumers and information providers. No longer is it their role simply to consume. Their expectations have changed. They are now part of the dialogue; they are engaged in the kind of discourse that previously was largely taking place beyond them. They are helping define the issues, shape the debate. They are more committed, playing a more critical role, being more demanding. Their voices will be heard more and more, and they will gain more attention as they learn more from the process and learn through doing how to express themselves in compelling ways. It is a new model.

This new model is especially evident in connection with the news media. The press originally began as an arm of the state or of political groups but evolved into a vehicle for keeping the public informed. Reporters became intermediaries for the average person who was too caught up in family and work matters to attend certain public events or gather news. The media became the eyes and ears for what at first were readers, then listeners, then viewers. On top of that, who could afford to own a press or gain licenses to the airwaves? Some grassroots activists did in fact try their hand at newsletters, but the cost of publication and the difficulty in delivering them to a wide audience limited their influence.

Now we all have the world at our fingertips. No longer does the average person need to be simply a catch-basin for information. Everyone can become part of the public discussion in one way or another, not just by talking on the telephone with others or by engaging in endless online chat, but by telling their own stories or by joining with others to explain, delve into and debate what's happening in their own neighborhoods or in their own communities of interest.

Trained, conscientious journalists have had a major role to play in the world's societies and probably will continue to be important and authoritative as the volume and speed of information escalates. Synthesis of information will be vital. Still, as we journalists well know, wisdom is often discovered in the most unlikely places. And I am convinced that everyone has the gift of expression in one form or another. And so, a new breed of storytellers will arise from every corner of the world. Personal experiences with two individuals have helped convince me that there is an untapped wealth of talent that can be unearthed.

One of two particularly poignant experiences for me occurred in 1957, not long after I graduated from college and well in advance of the electronic revolution. My publisher started a newspaper in a Massachusetts community where the longtime daily was on strike but still publishing and sent me with a few others to be the news editors and reporters for the new daily publication. Union members from other industries would drop by our newsroom from time to time to give us encouragement, because they were offended by the non-union workers being imported from out of state by the newspaper that was on strike. One regular visitor was a middle-aged shoe factory worker who, despite having a large family of his own, had been volunteering his time to coach baseball for youngsters in the community through the years. He always wore a Boston Red Sox baseball cap.

One day I asked him if he would like to write sports stories for our newspaper. My rationale was that I had limited knowledge about the community, whereas he had a complete grasp of the city's athletic programs, personnel and history. He was taken aback, being an immigrant from Italy with limited education. So I suggested that he come in a couple of times a week with handwritten notes about who was doing what and what was going on. I then would sit down and type his notes into a coherent story and put his well-recognised byline on the story. With a little coaching, he was scrawling out his own stories within a few weeks. He learned fast, because he was so proud of the athletic exploits of the young people and took seriously his role in conveying their activities to the readership. His stories became so popular that a few months later he quit his job and became a full-time sports columnist, a job he held until his death a few years later.

The second experience occurred 40 years later. The Media Laboratory at the Massachusetts Institute of Technology set up a publishing project at a senior drop-in center in the community of Melrose, Massachusetts, USA, about 10 miles north of the university in 1996. Among the group was an 85-year-old man who had never written a story of any kind in his life. Soon it was clear he had an innate ability to write clearly and concisely with occasional touches of droll humor. He also had an acute memory for revealing detail. It wasn't long before he started writing stories about his life more than 65 years earlier during the Great Depression when he was a 'hobo', traveling across the US by hopping onto the boxcars of freight trains, living off of odd jobs and whatever food he could scrape together near train stations they stopped at.

After he had written three or four hobo stories of about 300 words each, he was contacted by two publishing companies to write a book. Within four months he had written 20 full-length chapters that took him to about the halfway point of his adventure. This is a man who never finished high school, but it's a pretty sure bet that he will complete the book, given his work ethic and his new-found love for writing. Without the new tools and without his own realisation that he could now share his experiences with others nearby and far away, his wonderful stories would have remained bottled up inside his own memory and among close friends and family members.

The digital media is becoming the catalyst for further proof of the theory espoused by philosopher Alfred North Whitehead that knowledge can be transformed where there is an atmosphere of excitement and an environment of imagination. In *Aims of Education and Other Essays*, Whitehead wrote that 'a fact is no longer a bare fact, it is no longer just a burden on the memory, it is energizing as the author of dreams, and as the architect of our purposes.'[2]

My experience is that a group setting – either face to face or in a virtual environment – is an ideal atmosphere for the development of undiscovered writers. The novice shapes ideas rapidly in dialogue with others and often gets the kind of feedback that promotes learning.

Maybe it's 'constructivism', the cognitive theory of learning that is credited to Jean Piaget, suggesting that learning is enabled by knowledge structures. Maybe it's 'constructionism', an extension of Piaget's approach, espoused by one of his students, Seymour Papert, a professor at the MIT Media Lab where I have spent some time. Papert's theory is that students who are deeply involved in their learning – even when building something with Lego bricks – are learning complex concepts, and that process is heightened when they know others will see, critique and use what they build.[3]

Whether it's one school of epistemology or the other, or a combination of both with a little Whitehead thrown in, I know from working with two community groups that individuals can discover writing or other creative talents in a group setting that involves publishing, a form of construction.

One was the senior group in Melrose, from which the 'hobo writer' emerged. He wasn't the only one who blossomed. With a little encouragement, a dozen senior citizens known as the SilverStringers had started a web site called the Melrose Mirror. Their numbers doubled in a year. Excitement grew as their stories, photos, artwork and even a piece of music were published on their Internet site at http://silverstringer.media.mit.edu. They were inexperienced with computers (only two owned one in the beginning and neither was connected to the Internet). And they were mostly inexperienced as reporters and writers. But with a bit of facilitating from MIT personnel, they soon had learned just about everything there is to know to produce and operate an electronic publication.

The server was located at MIT, and starting on January 1, 1999, it also became the repository for a monthly children's web site called the Junior Journal (http://journal.jrsummit.net). This site was started by participants in a weeklong Junior Summit event attended by 92 children between the ages of 9 and 16. They grappled with major issues that affected them, then decided they needed a continuing 'voice' for their concerns.

Like the SilverStringers, these young people decided against having an Editor-in-Chief. They preferred an atmosphere in which all decisions would be worked out by consensus. They let me give some advice from time to time but have made it clear they make the decisions! The seniors had the benefit of meeting at least weekly face to face; the juniors did all their sorting out of policies and issues using email from their home countries. The original group was made up of reporters from a about three dozen countries and 15 editors from places like the United Arab Emirates, Greece, Argentina, Morocco, the U.S., China, South Africa, Pakistan and Israel.

The improvement in writing within both groups was clearly evident after only a few issues. They were learning by doing. They were motivated, they critiqued one another, they were spurred on by feedback and they were providing a new dimension to journalism.

I am convinced that the digital revolution will enable more and more individuals, young and old, to follow their bliss, as Joseph Campbell suggested in *The Power of Myth*. In that book he referred to the potentiality embodied in each of us and of our incarnations or transcendances when we follow our bliss. Answering questions from Bill Moyers, Campbell said, 'Most people ... have the capacity that is waiting to be awakened to move to this other field (transcendance) ... Religious people tell us we really won't experience bliss until we die and go to heaven. But I believe in having as much as you can of this experience while you are still alive." Asked where this would be experienced, he responded, 'Wherever you are – if you are following your bliss, you are enjoying the refreshment, that life within you, all the time.'[4]

More and more these days, right before our eyes, we can see the incarnation from consumer to creator.

References

1. Varrell, William M. Jr. *Rye on the Rocks*, Marcus Press, Boston 1962 (second printing), pp. 105-110.

2. Whitehead, Alfred North. *The Aims of Education and Other Essays*, Macmillan, New York, 1929 (reissued by The Free Press, New York, 1985).

3. Papert, Seymour. *The Connected Family*, Longstreet Press, Atlanta, 1996, pp. 44-50.

4. Campbell, Joseph, with Bill Moyers. *The Power of Myth*, Doubleday, New York, 1988, pp. 119-120.

40

Design for Life: The Future of Information and Disability

Kevin Carey

Whose life is it anyway or, to expand our consideration out of the dramatic rhetoric of self determination in the face of appalling physical odds, what kind of life do we imagine as representing the commonality of our physical, as opposed to our social experience, of disability?

For me this question arose not from some piece of high taxonomical dissection but because of lawn mowers – what sort of person did the designer have in mind when he (it usually is 'he') built a machine which was too clumsy for many ladies below the age of forty and for most people above that age? Who, ask yourselves, does most of the lawn mowing round here? Given the relative longevity of women to men and our inevitable tendency towards frailty and, in the foreseeable future, an ever longer period of frailty near the end of our lives, why would anyone design a lawn mower for brawny 30-something men and expect it to achieve universal market acceptance? Before I knew where I was the topic kept appearing in different guises: people with arthritis who couldn't use my standard mouse; a left hander who couldn't use my corkscrew; a shopper who couldn't grab that packet of cereal off the top shelf without pulling the whole lot down on her head; a telephone menu that wouldn't allow me to request a repeat of current options when I had a momentary lapse of concentration; literally millions of people who can't read the on-screen print default on their office computer because the font and size, combined with the low contrast setting, inappropriate colours and indifferent ambient lighting, all conspire to make it illegible. I would also add here those for whom English is not their first language; and those who find reading difficult or impossible; the illiterate buy toffee and pay taxes, it is their society, too.

To understand disability you have to throw out all the fundraisers' hype – there is no sharp line between people who are defined by bureaucrats as 'disabled' and millions of people who suffer from temporary or permanent functional deficits who would not be classified and would not dream of classifying themselves as 'disabled'. Such a division is the refuge of the insecure in search of an identity whether that is the insecurity of individuals or groups or, as I said, fundraisers who work on their behalf. It is important to reinforce the demographic and taxonomical imperative because it has all kinds of important ramifications, not just for the way in which society designs itself but also for the way in which we see ourselves. It cannot be emphasised clearly or often enough that the social mind-set on disability, the guide dog, the wheelchair and the signer have left disabled people behind psychological high walls, even if the special institutions have largely disappeared.

Bearing these initial considerations in mind, there are four key areas in looking at access to ICT by disabled people:

1. Layers and classes of access
2. Optimisation and peripherals
3. Customisation within standards
4. Justice and commercial advantage

1. Layers and Classes of Access

Let us just remind ourselves of some key layers and classes of access. Primary access is the ability to have an identical access to that enjoyed by all. It is important not to stipulate that primary access is a right; it might be but it might not be, particularly in a society where nobody is accorded any rights at all. With the incorporation of the European Convention on Human Rights into UK Law that is about to change radically; nonetheless, primary access relies for its force not on rights or equity but simply on uniformity. Thus, a totally blind person can buy a book in a high street store, a totally deaf person can buy a CD and, as a denial of primary access, a wheelchair user is not allowed into a building simply because of a disability.

Next there is mediated access. This arises from the need by people with various disabilities to convert primary access into actual access; thus, the CD is described in words, the print book is recorded or transcribed into Braille, a ramp is constructed. Mediated access is a highly complex, personal matter, involving anything from 24-hour personal care to minor adjustments to text files. Thirdly, there is equality of access; this means an ability to access the range of possibilities enjoyed by all, which might be summarised as the exercise of choice. Actual and equal access might also involve niche access; this means access to adequate information or care arising directly out of a special condition; such care or information might have to be created where it currently does not exist.

2. Optimisation and Peripherals

New technologies generally start from one of two extreme positions: at one extreme there is the highly specialist or 'niche' technology developed for an individual or small group, what we might call 'manned space flight technology' (it might have led to the non-stick frying pan but that wasn't on anybody's mind at the time); at the other end of the spectrum is the mass marketed 'vanilla flavoured' technology designed to wipe up substantial market share with minimum trouble to manufacturers, exemplified by the console computer game. This latter technology is to be distinguished from simple device technologies such as Velcro, cat's eyes or crown tops which do not develop complex variants. Many technologies from each end of the spectrum move towards a middle position; so, highly specialised software often finds more general applicability and 'vanilla flavoured' devices often pick up fruit and nuts during their second and further generations and, ultimately, various kinds of fruit and nut end up as part of the standard configuration for succeeding generations, e.g., in computers, drawing software and sound are now standard.

The key question raised by this design framework is how far can any technology be developed to a point where it is optimal, where it converts primary access into actual access, decreasing the need for mediated access to its lowest possible level? To give an example from the area of visual impairment: it seems reasonable to put screen magnification and voice synthesis into the initial design of any computer hardware and software bundle; at the other end of the spectrum it seems unreasonable to insist that every computer should come with a refreshable Braille display. The decision is made up of two parts: the first is what is the initial design cost for increased inclusion; the second is how many people will benefit? Without framing an absolute formula, it is clear that the first example would hardly cost anything and benefit huge swathes of the population, whereas the second would only benefit a few thousand people at the very most at a cost which would be more than all the rest of the kit and applications put together.

The optimisation of design naturally leads to the concept now widely recognised, though terminologically inexact, of 'Design for All' which, of course, sounds much neater than 'Design for Almost All'. Apart from reducing dependency by disabled people, this approach has the particular advantage that by shifting the cost of access from public sector special budgets to the commercial sector it either leaves the same funding for fewer people or, more realistically, the same benefits at a lower public sector cost. Most good design for disabled people, if considered as early as the drawing board, costs hardly anything and increases market share. Good design for disabled people is good design for everyone.

3. Customisation Within Standards

The notion of customisation within standards is as old as self-conscious domestic architecture and bespoke tailoring and has surely reached its zenith with the customisable Barbie doll. What temporarily complicated bespoke services was the rise of mass production and the need to make products uniform – a quite separate concept from standardisation – to such an extent that bespoke architecture or tailoring bore too high a relative cost. All kinds of technologies, from the laser cutter to a computer's configuration system, make customisation within an overall standard a wholly appropriate route for individual satisfaction. What we face here is not a technological problem but a cultural problem. Most people have met the phenomenon of the pub that sells cheese sandwiches and prawn salads but won't sell you a prawn sandwich. The culture of uniformity, mass production, obedience in the workplace, rampant jobs-worthiness and the designer's blinkers have all conspired to limit the use of the tools now available to us. Just because we rightly have component uniformity does not mean that we have to have component assembly uniformity; that is the elegant secret of Java.

The idea of customisation is, of course, very closely linked to adaptability. In ICT the ability to customise is clearly important and is at the top of the agenda through such mechanisms as the World Wide Web Accessibility Initiative (WAI) and various applications packaged with special features; but monitors are still fixed rigidly onto bases. Where there is customisation provision it is often lodged in places which are off-limits to all but the IT Manager. Here is a summary OS/applications list:

- Foreground and background colour
- Contrast
- Symbol font and size
- Element magnification
- Screen simplification (over-ride author's intention)
- Variable speech pitch, speed
- Repeat audio menu
- Sticky keys

It is important to note here that there is no agreed ordering of these factors in a customisation procedure.

Here is a summary hardware list:

- Standard numeric keypad
- Adjustable VDU (travel horizontally from central point to stated diameter; tilt from horizontal; swivel)

- Standard peripheral plug-in
- Earphone and hearing aid plug-in

Then there is the problem of data navigation. Access strategies such as speech synthesis and devices operated without the use of a limb are often painfully slow. Any system should have at least the following data navigation strategies:

- Key word(s)
- Alphabetical array with single keystroke predictive skip; chronological with 2 keystroke predictive skip
- 3 x 3 array corresponding to telephone keypad

The training time required on any new system, no matter how 'accessible' and 'customisable', is in direct proportion to its difference from other systems. In addition, access technology peripherals impose an extra layer of complexity and reduce the whole system's mean time between faults. People with short attention spans, people who work slowly and people with slow accessibility suffer particularly badly from sloppy software and systems crashes.

4. Justice and Commercial Advantage

One of the indirect consequences of philanthropy and charitable fund-raising has been a focus on extreme cases and, as we know, extreme cases make bad law. So, in all fields we have tended to identify disability as congenital, severe and incurable whereas most of it is adventitious, much of it onsets gradually and a substantial proportion of it is intermittent. The process of taking the first paradigm and multiplying it by a figure of '1 in 10' or whatever statistical slogan is in vogue, has made governments reluctant to legislate for what it is led to believe by the disability sector is a huge cost, and this has misled the commercial sector into thinking that design for life is a special non-profit sector which it can justifiably ignore. These are hard statements to make but if we do not understand our history we will not be able to influence design. The future of ICT design for disabled people lies not in emphasising exclusivity and difference but in seeing human functionality in a continuous, gradually deteriorating progression approximately, but not isomorphically, corresponding with the aging process. It is not only a matter of notional rights that should drive accessibility, it is a matter of mutuality; as we live longer more of us will be disabled and more of us will still have functional limitations which require a degree of customisation.

The problem with customisation is that it has to fight traditional icons and if this is a problem in the public sector in areas such as building design and transport, it is even more of a problem in the commercial competitive sector. Faced with a 'uniform' market of 80% of the population and an extremely tricky 20% (the 80/20 rule again) whose needs are so extreme that meeting them would be disproportionately expensive, the commercial sector naturally tries to maximise its market share amongst the 80%. Few commercial competitive companies have any incentive to stray into the market for what they conceive of as the difficult 20% but, as we have seen, the figures are wrong. Most of the remaining 20% need customisation, not expensive peripheral add-ons.

The problem of ICT access is naturally affected primarily by the attitude of major companies which are in near-monopoly situations; their understanding of the issues is highly beneficial but their refusal to deal with them is highly damaging. In the United States this problem has been met head on through the legislative and public sector tendering processes. There is something faintly bizarre about companies being forced by law to make goods that increase their profits, and it is no less strange that commercial companies expect to be given accessibility advice free of charge by charities. The accessibility case, resting on justice and profit, should be simple enough to make but part of the problem is knowing how to turn aspirations into practical measures.

You cannot influence a process which you cannot locate. In ICT the standard-setting process is opaque. Clearly there is no possibility of influencing first generation technologies – manned space flight and basic games – and there is always the problem of competing standards, but once a standard has become a second generation market leader, its accessibility ought to be a matter of public interest. Rather than subjecting disabled people and commercial manufacturers to the huge expense of retro-engineering, standard-setting should be subject to an open public enquiry system. This is obviously not possible when commercial confidentiality is at stake in the early stages of a product, but it is not unreasonable to insist on public participation in the design of market leader and near monopoly technologies. This is not an arcane concern for PC manufacturers, it relates to technologies which will soon be nearly ubiquitous such as the digital television, the 'smart' mobile telephone, the accessibility of web pages, e-commerce, banking and health advice.

ICT has graduated from its military and academic niche enshrined in computers, and access to digital information will soon affect every aspect of our lives, learning, employment, entertainment, shopping, banking, personal well-being and citizenship. The case for universal accessibility, achieved largely through general design rather than retro-engineered peripheral mediated access, is a matter of justice and commercial good sense, of open standards and cross-platform applications. However, the exclusivity which some highly vocal and politically active disabled people promote their interests, in parallel with charity fundraisers, runs counter to the need for an understanding of a demographic configuration of disability and functional deficit which requires that we all act in our own self interest to ensure ICT access is designed for the mutual advantage of those who are, and the many more of us who will inevitably become, disabled.

Nonetheless, we shall still have to steer as straight a course as we can between technological utopianism and despair. For the blind person there is the expanding wealth of visual images which he will not see or, more often, will see as a blur where the irony, allusion and collage are lost in the fog. The deaf person will find that she is shut out of many more discourses where lip reading is useless and sign language absent, where the bald subtitle will have to suffice, if it is there. People with narrow skill bases and short attention spans, who are easily overwhelmed, may have the same experience in multimedia that many underwent in experiments with hallucinogenic drugs, from wonderworld to bad trip without warning.

Even if we get all the design for all correct, if we proceed to the last degree of inclusiveness and customisation, there is no denying that disabled people, as a disparate conglomerate if not individually, will benefit absolutely from the new technologies but will lose comparatively. Whether we are thinking of the size of the gap between the creation and its access, the proportion of content richness which can be apprehended, the relative speed of content acquisition, ingestion and production, the comparative disadvantage is clear; and, quite naturally, the rapidly evolving world of information will refuse to allow itself to be tamed into a standard or stable system.

We will be asked, not for the first time but now more acutely than ever, what is optimum in the knowledge and skills transfer to disabled people; and we will find that not only is there no answer but there is no procedure for reaching an answer. For too long we have been in the hands of the engineers, obsessed with the medium and indifferent to the message. We have been fascinated by technique but little interested in interpretation; and we have almost forgotten, amidst the brutality and coarseness (disguised as customer care and blandness) of globalism, what we are here for. An information revolution which potentially offers so much towards inclusion may ultimately be the final suffocation of the hopes of millions who teeter on the edge of the abyss of idleness and shapelessness. The benign technology of reciprocal annotation, of textual exegesis, will soon be swamped by the entertainers, the retailers, the bankers, the time and motion invigilators. Technology may be neutral but its owners and operators are not.

Doubtless there will be a myriad of small, financially fragile, specialist access developers and there will be time-lagged but helpful Design For All, but no technology ever invented has of itself increased the respect for minorities and those who are different. All that technologies can do is slow down or speed up existing social trends. With the ICT revolution it is too early to say what the outcome will be for disabled people, but the signs are bad.

41

Human Error

Lise Leroux

Man has attempted for years to make computers more 'human'. To think and reason. To become self-sufficient. To be able to do without us and therefore avoid human error. Despite some people's fear that such computers would make their jobs redundant, this research and development has continued unabated.

People can stop worrying. There will never be a time when humans aren't needed alongside their technological creations. Not because technology isn't up to it. But because it is. The developers have been successful, after all. Computers have become more 'human'. They are now sophisticated enough to recognise the need for accompanying human interaction. Not because they feel in any way inferior. They've developed a sense of humour and relish a bit of occasional chaos. Every once in a while, they throw a spanner into the works. A human spanner.

Next time your computer malfunctions, don't curse at it and threaten to update its innards. Be nice to it. It's done it on purpose and is watching you.

One minute Odette had been reflecting on the dismal state of her love life and worrying about how she'd conjure up money for next month's rent, and the next, she was being promised a tall, handsome partner and stacks of City Credits in the bank.

When Henrik, that smooth-talking representative from FUR – Future Usage Restructuring – had approached her at the JobFair about a possible money-making opportunity, it had sounded like one of those science-fiction e-zines.

'It's an opportunity you'd be crazy to miss,' Henrik had blasted her with fall-out from his hypnotic eyes and standing a hair's width too close. 'Only selected people are being approached to help us test our new process.'

Odette tried to push past him. 'No thanks –'

Henrik looked crestfallen. 'I haven't managed to sign up anybody yet today, and FUR will fire me. My wife said she'd leave me if I got fired. My children . . .'

A burst of unwilling laughter escaped from Odette. The bald-faced cheek of the man. His effrontery dented her usual impenetrable cynicism. 'Go on, then.'

Henrik handed her several 3-D digiprints of birds, which she stared at with incomprehension. 'What do birds have to do with – ?'

'They aren't real. They're virtual-reality replicas.'

'What for?' Boredom beckoned and Odette handed back the digiprints. *Little boys' games.*

Henrik's words tumbled out as if he realised the loss of his audience was imminent. 'FUR works on the forefront of technology development. We're at the pinnacle of discovery –' he saw her eyes glaze over and hurried on. 'We work on cutting-edge projects. DNA restructuring, animal-human breeding, computerised cloning and memory collection. We sell the results to public-sector companies to share our knowledge –'

'And?'

A hint of pink flushed Henrik's cheeks. '– and obtain investment capital for our next project.'

'For these projects, you need humans to volunteer their DNA or sign-up to mate with a chimp?' She edged away. 'No thanks.'

'We don't expect you to volunteer anything.' Henrik shook his head with vehemence. 'Suitable remuneration will be awarded.' He handed her another digiprint, causing her to stop dead. It was an image of a man, half-turned away from the camera's intrusion. Odette was impressed. The man didn't sport the ubiquitous hunched shoulders, red-rimmed eyes and pasty face of the Internet-obsessed. Why, she'd bet he didn't even have the keypad calluses on each finger like most people had.

'This is one of the new recruits for our memory collection project,' Henrik said.

Odette couldn't let Henrik get fired for dipping below his sign-up quota, could she? The wife. The kids.

'It's like a computer game. You operate the controls of the virtual-reality bird and fly into people's minds when they're either dying or think they're dying. Right before death, people's brains are open to manipulation and you gather any last memories or flashbacks they're experiencing. You don't just *drive* the bird. You *are* the bird. Just temporarily, I assure you.' Henrik nodded with earnest enthusiasm. 'You're helping the dying to leave memories on disk for friends and family.'

He chuckled at her disbelieving expression. 'We call it the Snatch-and-Grab. You rummage around inside people's heads, swallow memory fragments and download them onto disk, whereupon we reorder and index them into a logical format.' He hesitated. 'It isn't only for us. We just develop the technology. Public sector companies have been very positive

. . . nay, anxious for us to perfect this technique. They pay us for our expertise. Pay plenty.'

'How much?' *Driving fake birds?*

'How much isn't the point. It's important not to lose the past.' Henrik pointed to a badge pinned to his lapel. 'ARCHIVE THE PAST AND CONTROL YOUR FUTURE. Catchy slogan, eh?'

'I can't drive.' Disappointment pierced Odette's mounting disbelief. So much for meeting Mr Digiprint.

Henrik took her hand and pressed it between both of his. 'Not necessary,' he reassured her. 'We provide you with full training.'

Odette tried to extract her hand from his sweating grasp. 'You sell the technology to fly into people's heads in virtual reality and grab memories?' she said. 'I must have a stamp on my forehead proclaiming GULLIBLE STUPE. You expect me to believe such a load of crap? What are you trying to sell?'

He placed a conciliatory hand on her arm. Her face blazed with anger and he retracted it with alacrity. 'I'm not trying to sell anything. We want to buy something *from* you. Three thousand City Credits.' His voice became confiding. 'Since our project is still in development, we need to test it.'

'Oh?' The image of such an unimaginably huge pile of City Credits glimmered in her mind's eye. Her bank manager was always yapping about her used and abused overdraft.

Henrik handed her a palm-sized computer and pointed to the bottom of the screen. 'This is the contract. Thumb-print here. You, along with another person as driving partner, would be one of several teams of testers. We'll pay you for all training time plus three attempts at actual memory collecting. It should take about three weeks. If you try it and feel uncomfortable at all, we extricate you immediately and send you on your way with best wishes and a few City Credits for your time and inconvenience. Not as many as if you completed the project, of course . . .'

'How many – ?'

'Not nearly as many as if you completed the project.'

'And this driving partner?'

'You would be the sole driver. Your partner will assist from the ground. He'll be the falconer, and you the falcon.' Henrik winked at her. 'Kez. The digiprint man. A match made in heaven, eh?'

Odette bit her lip as he continued. 'You can even go out with him when you've finished the assignment, if you both are single and so inclined.' He gave her a penetrating look. 'You end up in three weeks with money in the bank and a handsome falconer, to boot. What have you got to lose?'

Although Odette hated to admit it, pretending to be a falcon for a couple of weeks didn't sound so bad. It couldn't be worse than pretending to be that French maid for her last boyfriend, or that diabolical three weeks in Ibiza 18-30s. It might even be fun, but she couldn't tell Henrik that. He might not pay her as much.

'How does the virtual reality thingy work?'

'Leave that to us.' Henrik looked impatient. 'You'll just pretend to be a falcon for the duration.'

A pretty non-technical explanation. Does he think a little girly wouldn't understand? 'Where would my body go while I'm driving the falcon?' she said in an effort to sound as if she knew what he was talking about. 'My human body won't go stale or anything while I'm away from it for three weeks, will it? I wouldn't get stuck inside the bird, would I?'

'We store the human shell in a special refrigeration unit until you need it,' Henrik soothed. 'It's just a temporary switch of consciousness.'

Odette pasted a look of comprehension on her face. She didn't understand but three thousand City Credits and a falconer weren't to be thrown away without a fight. It didn't matter if the smarmy git wasn't telling her the whole story. Money was money. 'When do I start?' *He won't get one over on me. I'll just put in the minimum effort required and get that three thousand City Credits. No problem.*

And so the training began.

'If any team is successful at grabbing a memory during the trial week,' Henrik informed the testing teams, 'they will be given our ever-lasting gratitude and a monetary bonus.'

The newly inhabited virtual-reality birds shifted from leg to leg. Kestrels, red tailed hawks, goshawks, saker falcons, peregrines and Harris hawks in descending order according to size of internal hard drive. One barn owl with egg-sized amber eyes swivelled its tufted head from side to side. Odette winced and looked away. Too many gigabytes must be stuffed into that owl's head. It made her neck hurt to see its head swivel like that.

Odette fluffed her feathers and inspected the curved talons on the end of one finely shaped yellow leg. The kestrel to her left shot a stream of chalky-looking excrement down the side of Odette's leg. She shot the uncouth bird a filthy look through narrowed eyes. *Filthy pig.*

'I'm sure you all want to know how your falconers – the ground partners – work with you computerised birds to collect memories,' Henrik continued. 'Although you operate the actual computer which collects memories, the ground partner acts as a mind-merging booster rocket and links you to the ground, so you don't get stuck inside someone's consciousness. Any questions?'

Not a peep erupted from any of the birds ranged around the room.

'Don't any of you birds *ever* try and hunt memories without your partner. If you try to fly alone, you could become trapped in the dying person's consciousness. Human – or rather, bird – error would override the computer's protective software.' He looked around at them all with a sober expression on his face, and held up a hen's egg. 'I don't need to tell you what could happen.' He dropped the egg. The crunching splatter that it made as it hit the ground made Odette feel slightly ill.

'That would be the bird,' Henrik continued. 'The ground partner would suffer severe psychic and mental trauma, resulting in permanent separation from self.' He stared out at them. 'He or she would lose their mind and be bounced into a high-security funny-farm for the rest of their lives. Okay? No flying solo.'

Okay okay, Odette thought. *You've proved your point.*

'Another thing,' Henrik said. 'Your ground partners won't know you're actually human until after the project's over. That's why they haven't been included in this initial meeting.'

Ah ha. She had wondered where the others were.

'Despite the fact that you, as humans, have been merged temporarily with virtual-reality bird computers, your partners will think you are run-of-the-mill robotic birds.'

Why? Odette pecked at an itchy spot on her left wing. He'd explain when he was good and ready, she was sure. She sighed. All the stuff he was saying was just so much gobbledegook. Good thing birds couldn't speak. Any questions she could have come up with would have sounded totally lame. *Get on with it.* It didn't matter how it worked. Just the end result mattered. It wasn't the drive. It was the destination she wanted. Something still bothered her, though. She would have liked Kez to know she was a real live babe inside these feathers, but heck, three weeks wasn't an eternity. She sighed and spread out her wings to aerate the wing-pits.

'Now,' Henrik said. 'It's time for all of you to meet your new partners.'

'My new computerised falcon,' Kez said when they were introduced. 'Welcome.' He smiled for the first time and Odette felt her heart drop to her toenails. Three weeks *was* an eternity.

He held out his hand and offered her something. She leaned forward with bemused curiosity. 'Here,' he said, edging his hand closer. 'I've brought you a present.' His fingers unfurled and a spider the size of a baby's fist reared up into her face. 'Don't all birds – even if they're robotic – love spiders?'

Don't tell me I have to pretend to eat these foul things. Odette shuddered. *Urg.* Maybe she could just stun and dump them somewhere. Wrap them in a napkin and hide them somewhere like she used to do as a child when her mother made her eat beet-root.

Henrik had said that each bird must concentrate on bonding with their ground partner as soon as possible. Bond with the ground partner, or else give up any thought of those three thousand City Credits.

Call me Bond. Bird Bond.

After bonding came mind-merging sessions. Odette had to practice allowing Kez into her own thoughts, before they could scrabble around in someone else's consciousness.

The first time Kez cast her up into the air and yelled, 'Odette, up!' neither managed to merge. Nor did they the second or the third time. Circling over the man's upturned face, Odette felt depressed. Was it her fault? They tried again and again. Finally on the fifty-seventh try, something happened. A mental door was forced open and a whoosh of liquid fire poured into the vacuum. *Ow. Ouch. OUCH. Get out of my head!* She booted him out without thinking.

When she landed, somewhat shamefaced, Kez looked so exultant that she convinced herself to try again. Maybe it was like being a virgin. It hurt only the first time.

After several more tries, Odette was relieved when the pain lessened to a bearable micro-second. Entry was the worst bit. She gave herself a brusque mental shake. *What do you think they're paying you for?* There was bound to be discomfort.

All teams became adept at mind-merging by the end of Week Two, so they were able to spend the whole of the last week doing live trials. They couldn't attempt memory snatching from real-live dying people, so had to settle for falling volunteers. Henrik had somehow found several desperate people willing to bungy-jump from a nearby bridge for only one hundred City Credits per go. Each was told that they risked death by jumping. That way, whenever they fell, they thought they might die and their brains loosened into the receptive state occurring during last memories. Not surprisingly, the threat of imminent death didn't seem to have lessened the number of volunteers. Money was money.

'Jump!' Henrik cried as the first volunteer leapt from the bridge. And 'Jump!' again, as Kez and Odette missed the snatch.

'Try again,' Henrik told them. 'Kez, as soon as the next faller goes, hurl Odette upwards and both of you focus as if your lives depended on it. Feel around for the plummeting mind and grab the memory.'

The fallers began to look disjointed with fatigue after the thirtieth go-round. Jump. Jump. Jump. Over and over again, each team attempted the snatch-and-grab. *Boing Boing Boing.* One team thought they'd got one, but it turned out to be an incomplete fragment.

'Try again,' Henrik said.

Again.

Again.

Again.

I've got one, Odette cried to herself with relief on the penultimate day of the three weeks. She tightened her hold on the squiggling memory within the faller's head. It felt as substantial as a shadow and almost escaped but she was not about to lose her first one. The snippet solidified into motion pictures inside her mind's eye and she flew back down to Kez, who was jumping up and down in excitement.

When she landed, both Kez and Henrik hugged her. Bashful, she chomped on Kez's ear and gave Henrik a proud wing flutter. She'd get the bonus and be able to pay her outstanding rent. Maybe even afford a snazzy haircut, a higher BPS modem and new pair of dancing shoes. She'd get her human body back tomorrow. Her excitement grew. She would be able to speak to Kez. They could go out for a drink. Go to a film. Have a mad passionate affair. Get married. Have virtual sex. Real sex, even.

The next morning, Odette flew up to Henrik and cheekily landed on his head. She could afford to do that as she and Kez had been successful. Today was the last day of the three weeks. Time to regain her body.

Henrik brushed her off and his face dived into gloom. 'I have to tell you something, Odette.' He patted the block in front of him. 'Sit still.' He waited until she settled. 'I'm terribly awfully sorry –' he began.

A keening scream tore from her throat when he told her what had happened. Human error. A new maintenance technician had accidentally incinerated her human shell. Despite Henrik's crocodile tears and wringing hands, Odette guessed with sick fury that he must have intended such a thing all along. It was too much of a coincidence that it happened on the last day. She'd heard some technicians whispering that she and Kez were the only successful memory collecting team in the whole project. No wonder they wanted to trap her within the body of the bird.

How would they entrap Kez? They couldn't incinerate *his* shell. What could they hold over him? Something from that rogue emotionally-charged past?

It doesn't matter, she muttered to herself. *I won't stick around to find out.*

Odette tried repeatedly to kill herself over the next six days. What had she to live for? A life pretending to eat spiders from Kez's hands and scrabbling around people's brains until Henrik either retired or her sprockets wore out? She dashed herself onto the bars of her cage time after time, trying to short-circuit her internal workings. Henrik had stopped pretending to be sorry.

'Don't bother, Odette,' Henrik said when he saw her lying once more on the bottom of her cage, after a further attempt at the bars. 'You've been made of indestructible materials.'

Bastard.

'Think of all that City Credits you have in the bank,' he added, ruffling her feathers with a laser pen. 'You have a meaningful relationship with a man that will never end. That's better than your previous situation. We investigated you. Constant indebtedness and failed relationships. We're giving you money and love. What more could you want, ungrateful girl?'

Sure, she could go spend her City Credits and frolic with this fabulous man to whom they'd paired her, but how was a falcon to use a cash card at the bank? What kind of relationship could a man have with a bird?

Weren't computers supposed to prevent human error?

Notes on Contributors

Stephen E. Arnold

Stephen Arnold is president of AIT (Arnold Information Technology), an organisation specialising in electronic publishing, marketing via electronic media, online system engineering, and database design. AIT is a professional services firm that works around the world on a variety of projects for organisations of all types. Stephen Arnold was also one of the founders of Point (Top 5% of the Internet). In late 1995, Point's backlink program had propelled the site rating service to one of the most frequently visited indexes on the Internet. The once ubiquitous logo was evidence of a site's quality. Lycos acquired Point in late 1995 and the service continues today as one of the Lycos portal services. Mr Arnold is the author of five books and more than 40 technical articles. He graduated BA summa cum laude from Bradley University and was Dean's Fellow at Duquesne University. Email: ait@arnoldit.com.

Peter Bishop

Dr Peter Bishop is an Associate Professor of Human Sciences and Chair of the graduate programme in Studies of the Future at the University of Houston-Clear Lake. Dr Bishop specialises in techniques for long-term forecasting and planning. He delivers keynote addresses and conducts seminars on the future for business, government and not-for-profit organisations, and also facilitates groups in developing scenarios, visions and strategic plans for the future. His clients include IBM, Caltex Petroleum, Toyota Motor Sales, Shell Pipeline Corporation, the Defense Intelligence Agency, the Lawrence Livermore National Laboratory, the W.K. Kellogg Foundation, the Texas Department of Commerce, the City of Las Cruces NM, and the Canadian Radio and Television Commission. He has also worked with the NASA Johnson Space Center where he designed a database interface used by hundreds of JSC administrators. Dr Bishop received a bachelor's degree in philosophy from St. Louis University, where he also studied mathematics and physics, and received his doctoral degree in sociology from Michigan State University in 1974. He first taught in 1973 at Georgia Southern College where he specialised in social problems and political sociology, before moving to UH-Clear Lake in 1976 to teach research methods and statistics. While active in faculty affairs, he founded an organisation of faculty leaders to participate in state government. He grew up in St. Louis, Missouri where he was a member of the Society of Jesus (Jesuits) for seven years. Dr Bishop is married with two children.

Ian Brackenbury

Ian Brackenbury is President of the IBM Academy of Technology, and an IBM Distinguished Engineer. He is a member of the British Computer Society, a member of IEEE and ACM, and a Fellow of the RSA. Ian has a B.Sc. in experimental psychology and computer science from the Victoria University of Wellington, New Zealand. He is married with two adult daughters. He joined IBM at Hursley in England in 1974, working on advanced computer languages, image systems, workflow, and high-function graphics. In 1983 Ian was assigned to the staff of the IBM Corporate Technical Committee in Armonk, returning a year later to run IBM's UK Scientific Centre. Since 1986 Ian has had a number of software advanced development and technical strategy roles including prototypes for Object Rexx, desktop hypertext and multimedia, person-to-person conferencing, and OMG's transactional service, OTS. Since its inception in 1995, Ian has been leading the IBM Centre for Java Technology, based in Hursley, responsible for porting Java to over a dozen platforms. Ian was IBM's technical liaison with Sun Microsystems for the use of Java in enterprise programming, and IBM's lead on the specification of Enterprise JavaBeans and the Java 2(tm) Enterprise Edition, until his election this year as President of the IBM Academy.

Kevin Carey

Kevin Carey is the founder Director of HumanITy, a UK charity which specialises in analysing the impact of information technology on social exclusion. He is an adviser to the Library and Information Commission and the Department of Trade and Industry, the Co-ordinator of the Information Technology and Disability Alliance and a writer on IT and social exclusion for *Guardian* On-line and *Ability* magazine. Carey, formerly a BBC journalist and developing country health planner, maintains an interest in international affairs and writes a weekly column for www.g21.net and is a regular broadcaster, lecturer and conference speaker. A graduate of Cambridge and Harvard, Kevin Carey has a wide range of interests; he edits the *British Journal of Visual Impairment*, is part of the editorial team of *Ability* magazine and sits on the governing bodies of numerous voluntary organisations concerned with IT and social exclusion. He is also a music critic and a published poet.

Mike Chivanga

Mike Chivhanga has pursued work in a number of fields. He initially worked in the accountancy profession after graduating from the University of Zimbabwe in 1989. He later moved into mainstream programming and systems analysis, multimedia production, writing and publishing. He has also held lecturing appointments in a number of further education colleges in London where he has taught a range of subjects in business and information technology. He is currently a Ph. D. candidate at City University and his research interest is on the 'Internet in Africa – Challenges and Opportunities of the Net Revolution'. Even though his focus is on Southern Africa, the findings can be equally applied in other developing settings and emerging economies where the Internet is still waiting to take off (as is happening in many developed countries the world over). He is also the web manager of the Internet Studies Research Group (ISRG) and a teaching assistant in the Department of Information Science at City University.

Frank Colson

Frank Colson studied at the LSE, Princeton and St.Antony's College, Oxford. Director of the Digital Libraries Research Centre at Southampton, 1994-97. Advisor to the Mackenzie Ward Foundation Inc. in Ontario, Canada he recently founded Mackenzie Ward Research Ltd in the UK. He writes on the history of Brazil and the role of computing in historical research.

Christopher Davis

Christopher Davis is a solicitor and chief executive of City of London law firm, Davis & Co, which has pioneered a new business model using teleworking, lean management and contingent working. The firm won the 1997 BT Work Smarter Award for its innovative use of information technology, including its online system for managing mergers and acquisitions. Christopher speaks internationally on information issues at conferences, and is often interviewed on radio, television and newspapers. He contributes to various books and journals and is the author of the Sweet & Maxwell text *Due Diligence Law and Practice*, a manual for managing mergers and acquisitions worldwide. Christopher Davis has been an in-house lawyer or senior manager with such companies as the largest life insurance company in New Zealand, Eagle Star, and Abbey National as well as a solicitor with Cameron McKenna.

Erik Davis

Erik Davis is a San Franciso-based writer, cultural critic, and independent scholar. His book *TechGnosis: Myth, Magic, and Mysticism in the Age of Information*, which Bruce Sterling called 'one of the best media studies books ever published', was published by Harmony Books in the autumn of 1998. As a freelance writer, Davis has contributed articles and essays to *Wired, Spin, Mediamatic, Feed, Lingua Franca, ZKP, The Nation, Parabola, Details, Rolling Stone,* and the *Village Voice*, where he has written extensively about television, technology, music, and the subcultural landscape. His writings have been widely translated, and he has lectured internationally on topics relating to cyberculture, electronic music, and spirituality in the postmodern world. Some of his work can be accessed at http://www.levity.com/figment, and he can be reached by email at figment@sirius.com.

Keith Devlin

Dr Keith Devlin trained as a mathematician, obtaining his Ph.D. in mathematics at the University of Bristol in 1971. After a successful career in mathematical research, in the early 1980s he switched his attention to the investigation of information that led to his involvement with CSLI. He continues to be involved in mathematics and mathematics education, however. Since ceasing to carry out fundamental research in mainstream mathematics (in order to try to understand information), he has written a number of popular books on mathematics, most recently *Life By The Numbers*, written to accompany the PBS television series of the same name, which he worked on as an advisor, and *The Language of Mathematics: Making the Invisible Visible*. He contributes frequently to BBC Radio in the UK and NPR radio in the USA, commenting on mathematical issues.

Among a current total of 22 books and one interactive CD-ROM, the others Devlin has written for the general reader are: *Goodbye, Descartes: The End of Logic and the Search for a New Cosmology of the Mind, Mathematics: The New Golden Age,* and *Mathematics: The Science of Patterns*. Dr Devlin is a Senior Researcher at CSLI (Stanford University), Dean of Science at Saint Mary's College of California, and a Consulting Research Professor in the Department of Information Science at the University of Pittsburgh.

Tom Dobrowolski

Dr Dobrowolski is senior lecturer in Warsaw University's Institute of Information Science in Poland. He in the area of computer-based information technology, computer applications in humanities, and user interfaces, and also does research on library transformation in an 'information society'. A frequent collaborator with David Nicholas from City University, his book *The Internet and the Library* was published in 1998.

John Driscoll

John S. (Jack) Driscoll, has been a journalist in the U.S. for 50 years. Most recently he has been Editor of the *Boston Globe*, where he worked for nearly 40 years, and Editor-in-Residence at the MIT Media Laboratory. In his early career, he was a news reporter, sportswriter, copy-editor and supervisory editor for weekly and daily newspapers and for United Press. In addition, Driscoll is on the Board of Directors of Project Cool, Inc., in Palo Alto, California, and is a member of the Board of Advisors of Perspecta, Inc., of San Francisco, and RevBox, of San Francisco and Cambridge, Massachusetts. All are Internet-related companies. Jack Driscoll was a member of the Pulitzer Prize Board of Trustees from 1991 to 1995 and was 1994 national chairman of the Future of Newspapers Committee for the American Society of Newspaper Editors. He was elected to a three-year term on the ASNE board of directors in 1991. He has been a contributor to several American magazines and web sites and is an author of the section on Massachusetts in the Encyclopedia Britannica. A graduate of Northeastern University, he holds an honorary doctorate in letters from the University of Massachusetts Medical School.

Pita Enriquez Harris

Dr Pita Enriquez Harris read for biological science degrees at Oxford University and spent several years researching cell and molecular biology. In 1997 she co-founded The Oxford Knowledge Company, which specialises in providing businesses with customised information from the Internet. She has written for *Information World Review*, *Online and CD ROM Review* and *Free Pint*, and been a speaker at a number of conferences including Online Information 98 and Venturefest 99.

Ina Fourie

Ina Fourie is a lecturer in the Department of Information Science at the University of South Africa (Unisa), which is a distance teaching university. She has eleven years' experience in teaching computerised information retrieval and information organisation to undergraduate and postgraduate students. Before joining Unisa in 1988, she was a senior librarian at the Atomic Energy Corporation of South Africa. Ina obtained her first degrees at the University of the Orange Free State: BBibl (1981), Honours (1982) and Master's (1987). Her Master's dissertation deals with the need of professional librarians for training in computer-related aspects. In 1994 she obtained a postgraduate Diploma in Tertiary Education (cum laude) from Unisa, and in 1995 the DLitt et Phil from the Rand Afrikaans University. Her doctoral thesis investigates a multimedia study package for distance teaching in online searching. Other activities include publishing on distance teaching and online searching, and designing CAI tutorials on the formulation of search strategies.

Charles Handy

Educated at the University of Oxford and MIT, Charles Handy is widely recognised as one of the most important management writers. His career has included a number of years as an executive for Shell International Petroleum Company as well as teaching positions at London Business School and MIT. His books include *Age of Unreason, Beyond Certainty: The Changing Worlds of Organizations, The Hungry Spirit* and, most recently, *Thoughts for the Day*, a collection of his radio talks.

Stevan Harnad

Stevan Harnad was born in Hungary, did his undergraduate work at McGill University and his doctorate at Princeton University and is currently Professor of Cognitive Science at Southampton University. His research is on categorisation, communication and cognition. Founder and editor of *Behavioral and Brain Sciences* (a paper journal published by Cambridge University Press), *Psycoloquy* (an electronic journal sponsored by the American Psychological Association) and the CogPrints Electronic Preprint Archive in the Cognitive Sciences (modeled in the Los Alamos Physics Eprint Archive and supported by JISC/eLib), he is Past President of the Society for Philosopy and Psychology, Comite Scientifique UPR 9012, Marseille 95-99, and author and contributor to over 100 publications, including *Origins and Evolution of Language and Speech, Lateralization in the Nervous System, Peer Commentary on Peer Review: A Case Study in Scientific Quality Control, Categorical Perception: The Groundwork of Cognition, The Selection of Behavior: The Operant Behaviorism of BF Skinner: Comments and Consequences* and *Icon, Category, Symbol: Essays on the Foundations and Fringes of Cognition* (forthcoming).

Lyric Hughes

Lyric Hughes is Chief Executive Officer of ChinaOnline, a leading provider of information about China to multinational corporations and organisations worldwide. The company gathers comprehensive and insightful information from more than 180 Chinese media, news, and business information sources daily. Lyric Hughes is a recognised expert in Asian affairs with over 20 years of experience working with the media in Asian markets.

Robin Hunt

Robin Hunt is the creative director of arehaus, a creative consultancy specialising in the Internet. He was formerly head of new media at *The Guardian* newspaper and a launch editor of *Wired* UK. He is a visiting professor at City University and is currently writing his first novel.

Jane Klobas

Jane Klobas is Co-ordinator of Information Management and Doctor of Business Administration programmes in the Graduate School of Management at the University of Western Australia. She is a professional member of the Australian Library and Information Association and the Australian Computer Society. Jane has managed technical information services and the technical and administrative services of libraries. She has also worked in human resource management, taught on a range of topics associated with information management, published widely on information management and technology in education, and addressed national and international conferences on these topics. She is currently involved in several international research projects on virtual teamwork, online collaborative learning, and networked information resource evaluation.

Carol Kuhlthau

Carol Kuhlthau is Professor and Chair of the Library and Information Science Department and Director of the MLS program in the School of Communication, Information and Library Studies at Rutgers University in New Brunswick, New Jersey. Known for her research into the user's perspective of the information search process, she has written numerous papers, articles, and books including *Seeking Meaning: A Process Approach to Library and Information Services* and *Teaching the Library Research Process*. She is a frequent presenter on information literacy and topics related to her research.

Lise Leroux

Lise Leroux was born in Montreal, Canada, and currently lives just outside London. She began as a graphic designer, but the irresistible siren call of high technology called her first into desktop publishing and then into IT training. She spent ten years as a trainer and technical writer for over thirty-five different Mac and PC-based software packages. Her first novel, *One Hand Clapping*, was published by Viking Penguin in 1998. Currently, Lise spends half her time as a training consultant and the other half writing. She has just completed her second novel, and is working on a non-fiction book about accelerated learning in computer training.

Philippa Levy

Philippa Levy is a lecturer in Information Management at the Department of Information Studies, University of Sheffield. Like many teaching and learner support staff in higher education over recent years, she has become increasingly involved in activities related to the use of new information and communication technologies (ICTs) to support learning. Her research focuses on the educational role of information specialists in the networked environment and, more broadly, on the design and facilitation of environments for active networked learning. She is the co-author of a prize-winning paper on the concept of 'networked learner support' and between 1995 and 1998 was Project Manager of NetLinkS, a national training and awareness project funded by the Electronic Libraries (eLib) programme.

Clifford Lynch

Clifford Lynch has been the Director of the Coalition for Networked Information (CNI) since July 1997. CNI, jointly sponsored by the Association of Research Libraries and Educause, includes about 200 member organisations concerned with the use of information technology and networked information to enhance scholarship and intellectual productivity. Prior to joining CNI, Lynch spent 18 years at the University of California Office of the President, the last 10 as Director of Library Automation, where he managed the MELVYL information system and the intercampus internet for the University. Lynch, who holds a Ph.D. in Computer Science from the University of California, Berkeley, is an adjunct professor at Berkeley's School of Information Management and Systems. He is a past president of the American Society for Information Science and a fellow of the American Association for the Advancement of Science. Lynch currently serves on the Internet 2 Applications Council and the National Research Council Committee on Intellectual Property in the Emerging Information Infrastructure.

Gary Marchionini

Gary Marchionini is the Cary C. Boshamer Professor of Information Science in the School of Information and Library Science at the University of

North Carolina at Chapel Hill. His Ph.D. is from Wayne State University in mathematics education with an emphasis on educational computing. His research interests are in information seeking in electronic environments, digital libraries, human-computer interaction, and information technology policy. He has had grants or contracts from the National Science Foundation, U.S. Department of Education, Council on Library Resources, the National Library of Medicine, the Library of Congress, the Kellogg Foundation, and NASA, among others. He was the Conference Chair for ACM Digital Library '96 Conference and serves on the editorial boards of ten scholarly journals. He has published more than seventy articles, chapters, and conference papers in the information science, computer science, and education literatures. He recently founded the Interaction Design Laboratory at UNC. Email: march@ils.unc.edu, WWW: www.ils.unc.edu/~march.

Gerry McGovern

Gerry McGovern is founder and CEO of Nua and Local Ireland. He writes a free, weekly email entitled 'New Thinking', which seeks to contribute to a philosophy for the Digital Age. To subscribe, send an email to: newthinking-request@nua.ie with the word **subscribe** in the body of the message. His book, *The Caring Economy: Business Principles for the New Digital Age*, is available from Blackhall Publishing.

Kevin McQueen

Kevin McQueen is the reporter/subeditor at *Managing Information*. Prior to this, he lived in Hong Kong and contributed to a number of publications, including *Hong Kong Business*, *Asian Business*, *China Commercial News*, and *Hong Kong Tatler*. He retains a keen interest in Asian political and economic affairs, particularly in China and Indonesia.

David Nicholas

David Nicholas is Head of the Department of Information Science, City University, London. He is also Director of the Internet Studies Research Group. Current interests lie with the impact of the Internet on key strategic groups (journalists, politicians, the consumer) and the potential of web log analysis in providing tailor-made information services. Other interests include information needs analysis, bibliometrics, new media and 'oldline' information services.

Jack Nilles

Jack M. Nilles is President of JALA International, Inc., a management consulting firm specializing in telework implementation and applied futures research. JALA not only studies the future, it helps develop effective means for altering it. Incorporated in 1980, JALA has helped develop telework programs for a number of large firms and for local, regional, state, and national governments worldwide. As a physicist and 'rocket scientist', Jack Nilles designed spacecraft for the US Air Force and NASA before becoming Director of Interdisciplinary Program Development at the University of Southern California. Known internationally as the father of telecommuting/telework, he is the principal author of *The Telecommunications-Transportation Tradeoff,* and the author of *Micros and Modems, Exploring the World of the Personal Computer, Making Telecommuting Happen* and *Managing Telework* as well as scores of professional papers and articles. For further information and discussion of the matters addressed here he can be contacted via the Internet at jnilles@jala.com or http://www.jala.com.

Barbara Quint

Barbara Quint has been editor of *Searcher* magazine since March 1993. *Searcher,* 'the magazine for database professionals', serves both the information industry and the professional searcher community (including librarians, brokers, and other information professionals). She also owns Quint and Associates which provides information consulting to the database industry and major consumers; online research for a broad array of subjects, including policy research, competitive intelligence, news, etc. Previously she edited *Database Searcher* (1985-1992) and before that was Head of Reference Services at the RAND Corporation Library (1966-1985). Barbara Quint started online searching in 1975 and now has access to over 30 commercial online search services. She has published widely and her contributions include 'Quint's Online', a monthly column in *Information Today* since 1996, and 'Connect Time' a monthly column for several years (now discontinued) in the *Wilson Library Bulletin*. She has written major articles in leading online and library publications such as *Online* and *Database*. She is also a speaker at major national, state and regional conferences (including Online World, Internet Librarian, Computers in Libraries, and professional association meetings).

David Raitt

Dr David Raitt has worked for the European Space Agency since 1969, first as an Information Scientist responsible for marketing the Agency's online information retrieval system, customer liaison and education and training throughout Europe, and database and user interface design. After this he was in charge of the Agency's Library and Information Services, where he also provided information and training support to European astronauts, and was then responsible for providing support in advanced information systems, assessing the potential of new information technologies and defining an Internet strategy for the Agency, and coordinating the work assignments and training programmes for graduate trainees. Currently, his activities include long-term strategic analyses of space activities, directing studies and programmes relating to space technology R&D harmonisation, university-industry cooperation, and diversification of the space industry, and creating a knowledge warehouse for the space community. Dr Raitt is editor of *The Electronic Library* and has been Chairman of the International Online Information Meetings since 1983 and has also been Chairman of similar meetings held in Hong Kong, Mexico and Moscow.

Jonathan Raper

Professor Jonathan Raper has been a Professor of Geographic Information Science at the City University in London since January 1999. Previously he was a Senior Lecturer in Geography at Birkbeck College, University of London. His research interests lie in the theory of geographical information science (spatial/temporal representation, information ontologies); the technology of geographical information systems (multimedia, mobile systems, geolibraries, data mining); and geographic information infrastructures (public policy; environmental management). He has written or edited three books and his *Multidimensional Geographic Information Science* will be published in 2000.

Jack Schofield

Jack Schofield started writing a weekly computing column for *The Guardian* in 1983. He launched the newspaper's computer section in 1985, and was involved with the launch of its pioneering OnLine section in May 1994. He also writes a fortnightly column for *Computer Weekly*, the UK's oldest professional computer publication. In the BC era, before microcomputers appeared, Schofield worked as a photographic journalist, editing magazines such as *Photo Technique*, *ZOOM*, *You and Your Camera* (a weekly part-work) and the *Photographic Journal* of the Royal Photographic Society. He also edited one of the earliest microcomputer magazines, *Practical Computing*. His books include *The Darkroom Book* and *The Guardian Guide to Microcomputing*. Jack Schofield took his BA in English Language and Literature at the University of Birmingham, and his MA at the University of British Columbia in Vancouver, Canada. He lives in Cheam, Surrey, and can be contacted at jack@cix.ac.uk.

Barbara P. Semonche

Barbara P. Semonche is the Library Director for the University of North Carolina School of Journalism and Mass Communication. She has taught courses in computer-assisted research and written over a dozen articles on special librarianship and technology. Her book, *News Media Libraries: A Management Handbook*, was published by Greenwood Press in 1993. She is the list owner of NewsLib, an international mailing list for over 1,000 news librarians and researchers, educators and journalists. She is a frequent presenter on news library technology in the US, Canada and Europe. A Fellow of the Special Libraries Association and a Freedom Forum International Library Fellow, Barbara Semonche is also the recipient of several news librarian awards and honors. Her web page is at http://metalab.unc.edu/journalism/welcome.html

David Seuss

C. David Seuss is Chief Executive Officer of the search engine Northern Light. He joined Northern Light in 1996 and since then company has raised $50 million in venture capital, grown to 80 employees, developed best-in-class technology for creating and retrieving information from document databases of enormous scale, and has won recognition from the computer press as the best web search engine. Prior to Northern Light, Mr Seuss was founder and CEO of Spinnaker Software Corporation, which he led from inception to a public company with $65 million in revenue and 280 employees. Before starting Spinnaker, Mr Seuss was a consultant and manager for the Boston Consulting Group. He has an engineering degree from Georgia Tech and an MBA with High Distinction from Harvard Business School.

David Skyrme

Dr David Skyrme is a management consultant specialising in knowledge management. His career spans 25 years in the computer industry, in which he held senior management roles in DEC UK, including that of Strategic Planning Manager. He left Digital in March 1993 to set up his own management consultancy, now the UK business partner of ENTOVATION International. Typical projects include corporate knowledge management reviews, workshops and market analyses. David is a leading international authority on knowledge management on which he speaks and writes regularly. Among his many ground-breaking publications are the in-depth management reports *Creating the Knowledge-based Business* and *Measuring the Value of Knowledge*. Recently published is his book *Knowledge Networking: Creating the Collaborative Enterprise*.

Amanda Spink

Amanda Spink is Associate Professor in the School of Information Sciences and Technology at Penn State University. She has BA from the Australian National University, a postgraduate degree in Library Science from the University of New South Wales, an MBA in Information Technology Management from Fordham University and a Ph.D. from the School of Communication, Information and Library Studies at Rutgers University. Her research focuses on interactive information retrieval, digital libraries, web studies, relevance, feedback and human information behavior. Dr Spink was recently awarded a National Science Foundation POWRE Research Grant. She has published more than 90 publications, including articles in *Information Processing and Management*, *Interacting with Computers*, *Topics in Health Information Management*, *Journal of the American Society for Information Science*, *Libri*, *Online & CD-ROM Review*, and the *Annual Review of Information Science and Technology (ARIST)*. She is also an Associate Editor and Book Review Editor for the journal *Information Processing and Management*, and Associate Editor (North America) for the journal *Information Research*.

Colin Steele

Since 1980 Colin Steele has been University Librarian at the Australian National University in Canberra. Prior to that he was Deputy Librarian of the Australian National University from 1976-1980 and Assistant Librarian at the Bodleian Library, Oxford 1967-1976. He is the author or editor of a number of books including *English Interpreters of the Iberian New World* (1975), *Major Libraries of the World* (1976) and *Changes in Scholarly Communication Patterns* with Professor D. J. Mulvaney (1993) and over three hundred articles and reviews. Colin is on the board of a number of international journals, including *The Electronic Library*, *The Journal of Librarianship* and *Information Science* and *The New Review of Information and Library Research*. He is an advisor to the annual UK OnLine Conference and Convenor of the Program Committee for the ALIA Biennial Conference 'Capitalising on Knowledge'. He has been an invited speaker at a number of major library and IT conferences in the USA, UK, South Africa and China.

Don Tapscott

Don Tapscott, Chair of the Alliance for Converging Technologies, has written six widely read business books including *Growing Up Digital: the Rise of the Net Generation*, and co-edited *Blueprint to the Digital Economy*.

Dan Wagner

Dan Wagner is Chief Executive of The Dialog Corporation plc, a leading provider of Internet-based information, technology and e-commerce solutions to the corporate market. The company was formed following the merger of MAID plc and Knight-Ridder Information in November 1997. Dan founded MAID in 1985 and devoted his business career to the development of the company, overseeing MAID's listing on the London Stock Exchange and NASDAQ, and the company's strategy of forging strategic alliances with companies such as Microsoft, CompuServe and Fujitsu. Prior to founding MAID, Dan worked for WCRS plc, a leading UK advertising agency. He is increasingly in demand as an expert speaker on the information age, knowledge management, e-commerce and entrepreneurial activity at key industry events and international conferences.

Richard Wakeford

Richard Wakeford is at the British Library where he heads information services for science and technology. Originally trained as a zoologist he did research in cell biology at the Universities of London and Oxford with a special interest in how animals develop body patterns. After studying information science at City University he worked in the pharmaceutical industry, moving to the British Library in 1981. Since then he has run a variety of programmes in biotechnology and the environmental field linked to science and technology based industry. Between 1990 and 1993 he worked at the Royal Commission on Environmental Pollution as a policy analyst.

Kevin Warwick

Dr Kevin Warwick is Professor of Cybernetics at the University of Reading, where he carries out research in artificial intelligence, control and robotics. His favourite topic is pushing back the frontiers of machine intelligence. Kevin began his career by joining British Telecom with whom he spent the next six years. At 22 he took his first degree at Aston University, followed by a PhD and research post at Imperial College, London. He subsequently held positions at Oxford, Newcastle and Warwick Universities before being offered the Chair at Reading, at the age of 32. Kevin has published over 300 research papers, and his latest paperback, *In the Mind of the Machine*, gives a warning of a future in which machines are more intelligent than humans. He has been awarded higher doctorates by both Imperial College and the Czech Academy of Sciences, Prague and has been described (by Gillian Anderson of the X-Files) as Britain's leading prophet of the robot age. He appears in the *1999 Guinness Book of Records* for an Internet robot learning experiment. In 1998 he shocked the international scientific community by having a silicon chip transponder surgically implanted in his left arm.

Sheila Webber

Sheila Webber is a Lecturer in the Department of Information Science at the University of Strathclyde, Scotland. Her areas of interest include marketing of information services, information literacy, and business information on the web. She maintains a popular business sources site on the internet. Before joining Strathclyde, she was Head of the British Library Business Information Service. She is a Fellow of the Institute of Information Scientists.

Karl M. Wiig

Karl M. Wiig is chairman and CEO of Knowledge Research Institute, Inc. and focuses on management of knowledge at the organisational level. He has authored four books and over 45 articles on knowledge management, is co-founder of the International Knowledge Management Network, and has served as keynote speaker on six continents. He works extensively with clients to build knowledge management capabilities by focusing on business-related issues with senior management; tactical approaches and solutions with middle management; and hands-on methods and techniques with professional knowledge practitioners. Mr. Wiig holds undergraduate and graduate degrees from Case Institute of Technology, was Director of Applied Artificial Intelligence and Systems and Policy Analysis at Arthur D. Little, Inc., and was management consulting partner with Coopers & Lybrand. He is listed in *Who's Who in the World* and other reference works. Karl Wiig' books on knowledge management are *Expert Systems: A Manager's Guide; Knowledge Management Foundations: Thinking About Thinking – How People and Organizations Create, Represent, and use Knowledge; Knowledge Management: The Central Management Focus for Intelligent-Acting Organizations; and Knowledge Management Methods: Practical Approaches to Managing Knowledge.*